POND AND BROOK
A Guide to Nature in Freshwater Environments

Michael J. Caduto

Foreword by Donella H. Meadows
Illustrations by Joan Thomson

UNIVERSITY PRESS OF NEW ENGLAND
Hanover and London

To God
who is the source of this book
through the gifts of life, love, and grace

University Press of New England

Brandeis University	University of New Hampshire
Brown University	University of Rhode Island
Clark University	Tufts University
University of Connecticut	University of Vermont
Dartmouth College	Wesleyan University

Originally published by Prentice-Hall, Inc., Englewood Cliffs, N.J.
Copyright © 1985 by Prentice-Hall.
© 1990 by Michael J. Caduto
Photos by Cecil B. Hoisington copyright © 1984, 1990
Photos by Michael J. Caduto copyright © 1984, 1990
Cover photograph © by Cecil B. Hoisington

Printed in the United States of America

∞

Library of Congress Cataloging in Publication Data

Caduto, Michael J.
   Pond and brook : a guide to nature in freshwater environments /
Michael J. Caduto ; illustrations by Joan Thomson ; foreword by
Donella H. Meadows.
     p.   cm.
   Reprint. Originally published: Englewood Cliffs, N.J. :
Prentice-Hall, c1985.
   ISBN 0-87451-509-2
   1. Freshwater ecology.   I. Title.
[QH541.5.F7C33   1990]
574.5'2632—dc20   89-38056
                                                           CIP

5   4   3   2

CONTENTS

iii

PART TWO
THE STUDY OF STILL WATER ENVIRONMENTS
LENTIC ECOLOGY

PART THREE
THE FLOWING WATERS
LOTIC ECOLOGY

PART FOUR
A LOOK AT WETLAND ENVIRONMENTS

FOREWORD

Like most people, I live two kinds of life: one at work and one at home. In both of those lives I keep running into Water—not just the water that I personally need to stay alive, but Water as an issue, as a resource, as a problem, and as a tool that I need to understand in order to live both kinds of lives.

In my work life I am a scientist and teacher of resource management. I go to countries all over the world to help people figure out how to grow food, manage forests, mine metals, and build economies so that their population's needs can be met. I am especially concerned about how needs can be met *sustainably*, which means not just today and tomorrow but for years and generations in the future.

When you are involved, as I am, with the whole planet, all its people, and all its resources, you begin to notice the central role of Water.

The primary need of people, after air to breathe, is water to drink—clean, pure water. But about one in three people in the world have no access to clean water. The only water they can find makes them sick and sometimes kills them. It would not cost much to fix this problem; only about one-third the amount of money the world's people spend on cigarettes each year. The human race is not yet meeting its own need for clean water, but it could.

Even in places like the United States, where the problem of clean water for drinking has been solved, the use of water is not *sustainable*. In much of the West, farmers, households, and industries are drawing up groundwater faster than it is naturally replaced. The wells go deeper and deeper. Some day they will be dry, and no one has thought through what will then happen to the farming and the cities. That pumping-down of groundwater is not sustainable, and not necessary. Water can be used less wastefully.

On the whole planet about one-third of the surface water and a considerable amount of the groundwater is polluted to the point where it is no longer usable. Wildlife in that water is dying. That is also not sustainable, and not necessary, either. We can have active cities and industries without putting their wastes into water.

You can see that when you are worried, as I am, about the "Big Picture," about how nations could manage to use this beautiful planet sustainably to meet people's needs, you need to know something about the Big Picture of Water.

That is what Part One of this book is about.

My work life encompasses the whole world, but my home life is concerned with just sixty rocky acres in New England, where I have a small farm. Water is at the center of things here, too.

Through the middle of my farm runs a stream called Blow-Me-Down Brook. The brook is clean and fast-running; we catch trout in it. Beavers live there too, and for twelve years now the beavers and I have been in a battle over whether the brook should be turned into a swamp. The beavers favor the swamp. They make dams that slow the flow of water and flood my road and my garden and my sheep pasture. I break up the dams so the brook can run free.

This battle with the beavers will certainly go on as long as I have strength to wage it. The productivity of my farm depends on my staying ahead of the beavers.

In one corner of the sheep pasture there was once a low spot, a little swamp not made by the beavers, where I wanted a small pond so the sheep could drink and the geese and ducks could swim. One summer we asked the soil conservation service to help us make a pond there. Big machines moved in and scraped out a deep hole and spread the gouged-up dirt onto the pasture (raising it higher so the beavers had a harder time flooding it). When they were done the whole place looked lifeless, like a strip mine.

But the hole filled up with water, and life returned. Reeds sprang up on the bank, frogs appeared, and fish. Now, several years later, there is a beautiful pond there. All kinds of creatures have moved in and set up housekeeping. I have always wanted to understand more about them. How did they get there? Which ones help keep the pond clean? Which ones eat which other ones?

Chapter Three of this book tells me about ponds like mine. Chapter Five tells about running water, like Blow-Me-Down Brook. And Chapter Six tells about the swampy wetland that was there before the pond, and about the wetlands that the beavers are out there trying to create right now.

Human beings, and beavers, are always messing about with the world's flows of water. We deepen them, dam them, stock them with fish or harvest fish from them, pump them up, rechannel them, put them into pipes. To do these things wisely, to manage the resources of this world, or of one single farm, we need to know more about Water and all the creatures that live in and around it. As we learn to understand all the magnificent things, big and small, that are happening in and around the world's waters, we can begin to see our planet as the unique treasure it is.

DONELLA H. MEADOWS
Dartmouth College

PREFACE

Vast and intriguing are the many worlds of fresh water and the life therein. From the intricate geometrical designs of microscopic algae to the lonely call of a loon that echoes in the recesses of your soul, the world of fresh water is an endless source of new discoveries. Yet, to the student in the field, this fascinating array of plants and animals can be overwhelming. This book provides the knowledge and tips for beginning an exploration that, for many people, becomes a lifelong journey of discovering the mysteries of life in fresh waters. If you are a seasoned naturalist, the breadth and depth of material covered in the text, accompanying images, and suggested activities will deepen your understanding of particular organisms, their interrelationships and interactions with the physical environment. As you read, you will see the nature and spirit of fresh water come to life through stories, facts, activities, photographs, and vivid illustrations.

This is more than a list of freshwater plants and animals—it takes a holistic ecological view of the complex world of freshwater life. Whether pond or lake, stream, river, or wetland, there is a chapter that discusses life in that environment. You will learn about the living and nonliving components of fresh water and how these fit together to weave an ecological whole. Although this book, for educational purposes, looks at the worlds of fresh water as seemingly distinct ecological units, the reader should remember that there are many overlaps and shades of gray between ecosystems. Different names are sometimes used by ecologists to refer to what are, in fact, ecologically similar environments. Most ponds, for instance, are shallow enough to be designated as marshes by wetland ecologists.

You are encouraged to first read through the book in its entirety. Then use the techniques described at the end of appropriate chapters to explore those environments that you find to be most interesting and accessible. Many chapters take the form of a journey through the different habitats within each environment, discovering first the plants and then the animals found there. Although many of these plants and animals are found throughout North America, the book is especially relevant for use in the eastern states and provinces, with emphasis on the mid-Atlantic and north Atlantic regions.

Chapters 1 and 2 introduce you to the unique properties of water and the ecological principles that are basic to understanding aquatic life. The effects of human activities on freshwater environments is also explored in Chapter 2. With this conceptual framework you have a reference point from which to experience the lives and living conditions of plants and animals in the environments described in

Chapters 3 through 6: ponds, lakes, streams, rivers, and wetlands. An inner journey awaits you in the Afterword, which uses stories, poems, and imagery to explore water as sage, mentor, and kin to the human spirit.

Unique in its scope and approach, *Pond and Brook: A Guide to Nature in Freshwater Environments* investigates all common freshwater environments and basic ecological understandings, from wetlands and deep lakes to vernal ponds. The writing style puts you into the out-of-doors with a "here and now" approach consisting of a liberal use of stories, beautiful illustrations, and timely photographs. The discussion of pollution and its effects will enhance your understanding of natural environments by describing the ways that human activities alter the delicate fabric of life. Teachers will find this to be a useful and comprehensive text for introductory-level courses in freshwater ecology at the upper high school and undergraduate level.

Use this book as a reference point from which to branch into more specific studies of particular freshwater environments, and as a catalyst for your own first-hand explorations. Let it guide you through the worlds of fresh water and show you how the fate of these worlds rests ultimately in human hands.

Michael J. Caduto is an associate of the Atlantic Center for the Environment and the Vermont Institute of Natural Science, a consultant for the Vermont Department of Education, and has served as adjunct instructor in environmental studies for numerous colleges and universities. His writing has appeared in *The Nature of Things* by Will Curtis, *Hands-On Nature* by the Vermont Institute of Natural Science, and he has had over 100 articles published in national and international magazines. He is co-author (with Joseph Bruchac) of *Keepers of the Earth: Native American Stories and Environmental Activities for Children* (Fulcrum, 1988).

ACKNOWLEDGMENTS

The writing of a generalist is, in some respects, a collective enterprise. I am grateful to the many people who were willing to spend their scarce time and energy to review and comment upon the manuscript for this book: Dr. G. Winfield Fairchild of West Chester University, Pennsylvania; Dr. Carol Folt of Dartmouth College, New Hampshire; Dr. William T. Fox of Williams College, Massachusetts; Dr. John J. Gilbert of Dartmouth College; Dr. Frank F. Hooper of the University of Michigan; Robert E. Richardson of the Division of Water Resources, Rhode Island Department of Environmental Management; Charles E. Roth of the Massachusetts Audubon Society; Dorion Sagan; and Ralph D. Scott of the Massachusetts Audubon Society. The efforts of Ralph W. Tiner, Jr., of the United States Fish and Wildlife Service, who helped me locate several obscure photographs, are greatly appreciated. Jerry Fish was instrumental in preparing the photographs for publication. Eugenia S. Marks, editor for the Audubon Society of Rhode Island, lent support with the logistics for fieldwork. A special mention for Alfred L. Hawkes, executive director of the Audubon Society of Rhode Island, for his important role in helping me get started teaching about the environment some years ago. And thank you to Dr. Francis C. Golet of the University of Rhode Island for his thorough review of Chapter 6, and for introducing me to the many wonders of those environments.

Throughout the preparation of this book, Mary Kennan, Cyndy Rymer, and the staff at Prentice-Hall, Inc., provided the expert guidance needed to bring this project to fruition. Joan Thomson, whose beautiful line drawings complement the text, persevered under the pressure of deadlines. A fellow aficionado of freshwater life and professional photographer, Cecil B. Hoisington, generously provided some of the photographs found herein. I can never say enough how much I appreciate my friends and family, who have helped me to keep together, body and soul, as the work progressed.

❧ PERMISSIONS ❧

Permission to reprint the following is gratefully acknowledged:

The drawing on page 13 from *Ecology and Field Biology*, 3rd ed., by Robert Leo Smith is redrawn by permission of Harper & Row, Publishers, Inc. Copyright © 1980 by Robert Leo Smith.

Excerpts from articles by Michael J. Caduto found in the *Report* of the Audubon Society of Rhode Island on pages 21–22, paraphrased material in Chapter 5, and the photo on page 210 are used with permission of the Society.

The chart on page 32 and text on page 123 were adapted from Charles R. Goldman and Alexander J. Horne's *Limnology* (1983) and are reprinted with permission of McGraw-Hill Book Co.

The photograph on page 44 by Bill Azano is reprinted with his permission.

Illustrations on pages 102 and 173 were adapted from Robert W. Pennak's *Freshwater Invertebrates of the United States* (1978) and are reprinted with permission of John Wiley and Sons, Inc.

The illustration on page 127 was redrawn from H.B.N. Hynes' *The Ecology of Running* (1970) and is reprinted with permission of the Liverpool University Press.

Excerpts from "Of Streams and Insect Life" by Michael J. Caduto which originally appeared in *Vermont Natural History* (1982) are reprinted with permission of the Vermont Institute of Natural History.

The illustration on page 171 was redrawn from W. T. Edmondson's (ed.) *Freshwater Biology* (1959) and is reprinted with permission of John Wiley and Sons, Inc.

Excerpts from "Wetlands, Our Great Providers" by Michael J. Caduto on pages 185–186 which originally appeared in Rhode Island Audubon *Report* (September, 1978) are reprinted with the permission of the Audubon Society of Rhode Island.

The photograph on page 199 from P. V. Glob's *The Bog People: Iron Age Man Preserved*. Copyright © 1965 by P. V. Glob. English translation © Faber and Faber Limited, 1969. Used by permission of Cornell University Press. Faber and Faber Limited, Publishers, and P. V. Glob, Director of the Danish National Museum, Denmark.

The photograph on page 207 is used with the permission of the United States Fish and Wildlife Service.

Marsh Gas Ballad from the article "Blue-Burning Swamp Bubbles" by Michael J. Caduto which originally appeared in the *Vermont Standard*, July 21, 1983 is used with permission of the Vermont Institute of Natural Science.

The poems on pages 222, 225, and 228 by Michael J. Caduto which appeared in *Vermont Natural History* (1983) are reprinted with permission of the Vermont Institute of Natural Science.

The quotes on page 225 from Bliss Perry's *The Heart of Emerson's Journals* (1939) are reprinted with permission of Dover Publications, Inc.

The quotes on page 225 from Odell Shepard's *The Heart of Thoreau's Journals* (1961) are reprinted with permission of Dover Publications, Inc.

The poems on pages 3 and 228 by Lao Tzu, and on page 225 (author unknown), are from David Brower's (ed.) *Of All Things Most Yielding* and are reprinted with permission from Friends of the Earth.

The paraphrasing of the story of Tollund Man on page 199 from P. V. Glob's *The Bog People: Iron Age Man Preserved* (1969) is reprinted by permission of Faber and Faber Limited, London.

The definition of wetlands on page 186 from L. M. Cowardin et al.'s *Classification of Wetlands and Deepwater Habitats of the United States* (1979) is reprinted with permission from the United States Fish and Wildlife Service.

The definition of ecological pollution on page 36 is used with permission from Dr. Frank F. Hooper, University of Michigan School of Natural Resources, Ann Arbor.

The photograph on page 19 from *National Water Summary 1983—Hydrologic Events and Issues* is used with permission from the United States Department of the Interior Geological Survey.

PART ONE

A WORLD OF WATER

CHAPTER ONE

THE WAYS AND FORMS OF FRESH WATER

What is of all things most yielding,
Can overcome that which is most hard,
Being substanceless, it can enter in
even where there is no crevice.

That is how I know the value
of action which is actionless.

LAO TZU, 5th century B.C.

One season it is a soft, billowing cloud sailing on warm summer breezes, and the next it is a hard coat of ice encrusting the shore of a pond. Sometimes it appears in a flash as a deadly, rushing wall that can roll hundred-ton boulders along a riverbed as though they were grains of sand. Or it can play softly the flutelike tones of a gentle stream spilling into a grotto of rock—a mountain stream. Water is at once a liquid that seems to meld graciously with the shapes of all that it touches, yet it has the strongest of wills; given time, few substances can resist its sculpting, dissolving action. In water, nature has found its chameleon guise, an expression of its every mood.

FIGURE 1–1: (*Photo by Cecil B. Hoisington.*)

As seemingly abundant as it is ever-changing—from tundra to tepid tropical pool, from sea to hail and rain—water is everywhere. It is a gift to all living things. Almost three quarters of the earth's surface is covered by water, and roughly 80 percent of our bodies are composed of water as well. The chemical makeup of our cellular fluids is thought to be like that of the ancient seas in which primitive life forms may have first appeared. Life would be impossible without the provenant waters. If any healthy person were placed in a room without nourishment, the ensuing thirst would be the first life-threatening force. Everything that nourishes our cells—nutrients, oxygen, organic compounds—all must arrive via our arteries and veins. Nor have other land animals or plants escaped completely from water. A growing plant needs 100 pounds of water for each pound of new growth. Without water, all plants and animals would be sere and lifeless. Water is truly the seminal element.

Fresh water, however, is not as ubiquitous as it might at first seem. When one stands at the shore of a large lake or listens to the rushing flow of a river, it is easy to imagine that the supply of fresh water is boundless. Yet the salty oceans contain 97 percent of the world's total water. In fact, if all the liquid fresh water in the world was gathered in one enormous container, its volume would be less than 1 percent of that contained in the oceans and glaciers. This includes all the fresh water found both above and below ground. Glaciers and ice sheets comprise about 75 percent of the total fresh water on earth, although, being ice, this water cannot support life. Much of our inland fresh water is held in a few of the largest lakes in the world, such as Lake Baikal in Russia, the Great Lakes in the midwestern United States, and Lake Tanganyika in eastern Africa. The balance of our freshwater supply is spottily distributed over the rest of the earth's land mass.

Vast regions of the earth are periodically or permanently deprived of water's life-giving force. During the dust bowl period of 1933–1936, when an intense, prolonged drought befell the fallow fields of the Great Plains in the United States, the sky was obscured by precious, airborne topsoil as it rolled over the plains like a blanket. Some homes were completely buried in rich, windblown earth, and fertile fields were reduced to hills and craters veined with deeply eroded gullies. Even our modern deserts, like the Sahara of northern Africa and those of the Middle East, with their endless expanses of shifting sands, are believed to have been rich forests of ferns and club mosses during the Carboniferous period 350 million years ago. From the remains of these forests were formed the fossil fuels that now power much of our industry and transportation.

❧GLACIERS: WATER AS SCULPTOR❧

When water is abundant, its power can astound. The slowly moving glaciers show how ice—composed of the same substance that will, as a liquid, conform to the shape of any vessel into which it is poured—has irrevocably altered large portions of the northern hemisphere, the Antarctic continent, and all of the world's highest mountain ranges. During the Pleistocene geological period, glaciers molded and shaped the terrain of most northern regions of North America and Eurasia. Massive sheets of ice and snow up to two miles (3.2 kilometers) thick, they carried house-sized boulders as if they were pebbles. During the peak of the glacial advance they covered what is now Canada and the northern third of the United States, overtopping even Mount Washington, the highest peak in the northeastern United States, which looks down over northern New England from a height of 6288 feet (1917 meters). Twenty thousand years ago, at the height of the Wisconsin glacial period, about one third of the earth's land mass was covered with ice. Sea level was as much as 350 to 400 feet (107 to 122 meters) below its present height. During the current, warmer interglacial period, the glaciers cover about 10 percent of the land above sea level. (See Fig. 1–2, p. 6)

FIGURE 1–2: Glaciers have sculpted the northern landscape.
(*Photo by Norman Herkenham, courtesy National Park Service,
United States Department of the Interior.*)

As snow and ice accumulated during the glacial epochs, their massive
weight depressed the lower layers outward from the edges. Friction at the
interface of land and glacier, along with the heat of the earth itself, kept the
lower layer of the glacier partially melted. This softer layer helped the
glacier to glide along its course. Grinding and rounding off as it went, the
continental glacier contained billions of tons of rock, gravel, and sand.
Under its enormous weight, the crust of the earth was driven down as far as
500–600 feet (152–183 meters) in some places. In fact, until the land
underlying the Great Lakes Region rose as the glaciers melted, in a process
called *isostatic rebound*, the lakes drained to the north instead of in their
current southerly drainage pattern.

Gradually, and unevenly, the Wisconsin glacial period, which was the
last of several episodes of glacial advances, came to a close around 10,000
years ago. Sometimes warming and melting the glacier back, sometimes
cooling and causing a temporary halt or minor advance, the warming trend
left a barren, rocky landscape strewn with glacial debris in front of the

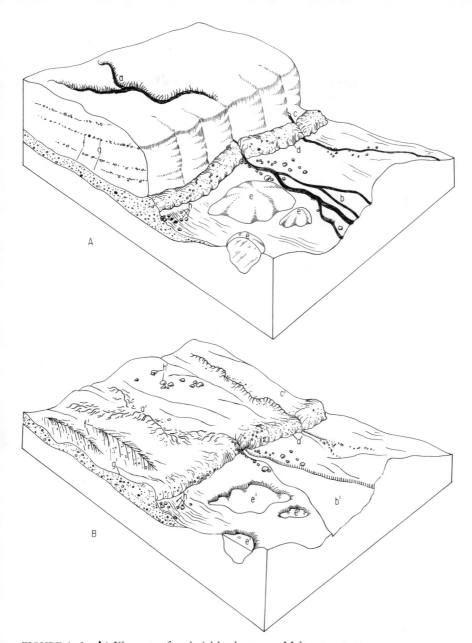

FIGURE 1–3: **A**) Elements of a glacial landscape. *a*. Melt-water stream over
the glacier's surface. *b*. Melt-water delta in front of the glacier. *c*. Melt-
water stream flowing from beneath the glacier. *d*. Debris deposited at the
glacier's edge. *e*. Ice blocks stranded in sediments in front of the glacier.
f. and *g*. Debris trapped within and beneath the glacier. *h*. Large boulders
transported and deposited by the glacier. **B**) Land forms resulting once the glacier
in Figure 1–3A has melted. a') Kame deposit. b') Outwash plain. c') Esker.
d') End moraine. e') Kettle holes (kettle lakes shown here). f') Drumlins.
g') Ground moraine. h') Glacial erratics.

7

glacier. Where melt-water streams had created snaking beds of sand and gravel on top of the glacier, *kame deposits* were left behind when the glacier melted. Eventually these streams flowed down off the glacier; as the waters slowed, they deposited their load in the lowlands, resulting in expansive *outwash plains.* Some melt water carved tunnels through the ice underneath the glacier and deposited sediments which later left serpentine rises known as *eskers.* Since these frigid waters deposited smaller particles as they began to slow, layers of larger and smaller particles were laid down according to the speed of the glacial streams. These banks of sand and gravel of glacial origin are known today as *stratified drift.*

Earth movers par excellence, the Pleistocene glaciers also left their mark on the landscape in the form of jumbled, unsorted heaps of rock, sand, and gravel called *till.* Hills called *moraines* were left where the glacier had dropped debris along its margin during the periods when its rate of melting and advance were equal, causing a stable front. Some examples of *end moraine,* that which marks the extreme southernmost point of the glacial advance, are the east coast islands of Nantucket, Martha's Vineyard, and Long Island. As the glacier melted and the sea level rose, these points of land were high enough to remain above water. *Recessional moraine* formed further to the north, where a minor cooling period temporarily created a stationary glacial front. As the immense load of glacial debris was dropped during times of rapid retreat, large areas were left with gently rolling sags and swells known as *ground moraine.* Some till was shaped by the glacier's movement into long mounds or *drumlins,* which are oriented in the direction of glacial flow. Frequently, elephantine boulders or *erratics* were carried and abandoned far south of their origin. Giant chunks of ice, too, were left in isolated pockets in the ground, buried under tons of earth that acted as insulation. As these ice blocks slowly melted, the overlying debris slumped into the space left behind and formed rounded depressions known as *kettle holes,* many of which have filled to form lakes and ponds.

CHEMICAL AND PHYSICAL PROPERTIES: THE MINUTE MYSTERIES OF WATER•

A glacier's gargantuan force is only possible because of the unique chemical and physical properties of the water molecules themselves, which account for water's baffling behavior and ever-changing appearance.

Ice floats. Freezing water expands with such a force that it was once used in quarries as a slow-motion dynamite, "blasting" tons of rock off sheer granite faces as it froze and exerted pressure on the sides of the holes into which it was poured. It takes such a great amount of heat gain or loss to change water's temperature that large tubs of it were once placed in underground root cellars to moderate the air temperatures. Water can look blue or green, yellowish or brown. And at any given moment, water can be seen as a liquid, solid, or gas somewhere in the world.

What is the nature of this substance that accounts for its remarkable qualities? Some of the answers can be found by looking at water's chemical composition. The familiar chemical formula for water, H_2O, indicates that each water molecule is comprised of two atoms of hydrogen and one of oxygen. *Covalent bonds* join the atoms within each molecule. Because the bonds within water molecules are uneven, the hydrogen atoms are left with a weak positive charge and the oxygen atoms are slightly negative, a situation that causes these atoms to be attracted to the atoms of neighboring molecules. As a result, hydrogen and oxygen atoms are quickly and incessantly forming and breaking weak *hydrogen bonds* with nearby oxygen and hydrogen atoms, respectively. It is this unique chemical structure that accounts for water's properties as a fluid material, its marked viscosity or fluid drag, and its *surface tension,* which creates the *surface film* that supports an entire community of plants and animals called the *neuston.*

Water is a liquid compound that can be frozen into a crystalline structure like a rock. Water molecules pack more densely as it cools, causing it to become heavier and sink. But at 39.2°F (4°C) water reaches its highest

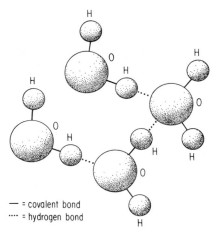

— = covalent bond
···· = hydrogen bond

FIGURE 1–4: Water's molecular structure. The asymmetrical covalent bonds formed in each water molecule result in a weak negative charge near the oxygen atom and a weak positive charge near the hydrogen atoms. Thus the oxygen and hydrogen atoms are attracted to the hydrogen and oxygen atoms, respectively, in neighboring water molecules, with which weak hydrogen bonds are formed. (Opposite charges attract.) Water's unique properties are a result of this chemical structure.

FIGURE 1–5: In ice lies strength and beauty.
(*Photo by Cecil B. Hoisington.*)

density, and as it cools further toward the freezing point at sea level of 32°F
(0°C), its molecular structure gradually expands, becomes less dense, and is
transformed into a highly organized crystal lattice. For this reason, ice is
lighter than the slightly warmer water beneath it, and so ice floats. If it were
not for this unusual property, ponds and lakes would freeze from the bottom
up instead of from the top down, creating a deadly circumstance for aquatic
plants and animals.

While these assertions are true for fresh water in a fairly pure state,
they will vary slightly if the water contains salts or other elements. The
freezing point of salt water is slightly below that of fresh water, depending
on the concentration of salt. This is why ocean water does not freeze until it
reaches temperatures below 28.4°F (−2°C).

FIGURE 1–6: Refraction causes submerged objects to appear to be in a different place than they really are.

Although the global temperature range of water is wide, from 27.1°F (−2.7°C) in the polar regions to 86.0°F (30°C) in our warmest seas, it is not nearly as wide as that on land, which has reached as low as −126°F (−87.7°C) in Antarctica and up to 136°F (57.7°C) in the Sahara Desert. Water temperature responds slowly to a rise or fall in the temperature of its surroundings. Its specific heat of one, or the amount of heat it takes to raise the temperature of a unit weight of a substance 1°C, is among the highest of all known substances. Water's latent heat of fusion is also high, which measures the amount of heat needed to melt one gram of ice without raising its temperature; likewise water's latent heat of vaporization, or the amount of heat needed to evaporate one gram of water without changing its temperature. For this reason, a large body of water such as an ocean, lake, or river is a natural hot-and-cold sink that moderates the temperature of the surrounding area, keeping the local climate warmer longer into the fall and cooler well into spring.

As we would expect, light also acts differently in the water than in the air. Suppose you were spear fishing as an American Indian would have done several thousand years ago. With muffled footsteps, you slowly and quietly make your way to the edge of a large pool in a stream. Once there, you assume a stalking stance, ready to thrust your stone-tipped spear when a fish happens by. In time, a large trout eases into the pool and swims lazily to within your range. Patiently, you wait for just the right moment to strike. Then—whoosh! The spear penetrates the water exactly where you saw the fish. But your shot has gone high of its mark! Why? Because light is refracted or bent as it reflects off the fish toward your eye, and the fish appeared to be at a different spot than it really was. Likewise, a fish, when about to leap at an intended prey flying over the water, must aim in front of its would-be meal.

A closer look at water and light can tell us why some water appears blue, some greenish, while still other water can look deep brown. Part of the sunlight that reaches water is reflected; this is the light we see. The proportion of reflected light to the total incoming light is called the *albedo*. White light is composed of all the colors of the spectrum: red, orange, yellow, green, blue, indigo, and violet. This is broken up as it enters the water, which acts like a giant prism. Some colors, such as green, tend to penetrate deeply, whereas red is absorbed so rapidly by water molecules and algae that it does not penetrate well. Blue is scattered and reflected by the water; so clear, clean water appears blue. Nevertheless, water does not always look blue. The surface can reflect gray clouds or green leaves overhead; the bottom sediment in the shallows of a lake or pond also impart color to the water. Dissolved organic matter and particles can give water a greenish tinge, as can the algal blooms of mid to late summer. Large amounts of organic matter in the water, such as that associated with bogs and swamps, can turn water a tealike brown. The time of day can also affect color, since more light is reflected during the morning and evening hours, when sunlight is striking at a greater angle.

❧ THE WATER CYCLE ❧

Gazing out over the blue waters of a lake on a summer day, you may begin to wonder, "What happens to the sun's energy when it is absorbed by the water? Where does it all go?" Stoop down and run your fingers through the water and you will feel some of that invisible energy in the form of heat. The sun's energy also causes evaporation, and creates clouds that carry water around the globe. The rain and snow fall down upon the highest peaks of Mount Everest and into the lowest valleys, replenishing the stores of fresh water on land, which run incessantly down into the oceans.

In the next glass of water you drink could be some water molecules that were once used to bathe children along the Tigris River of Babylonia during the reign of King Nebuchadnezzar, six hundred years before the birth of Christ. Our planet's supply of water is finite, and it is constantly circulating through the atmosphere and falling to earth. Sunlight and wind cause water to evaporate into its gaseous state and lift this water vapor up into the atmosphere on the warm air currents that rise from the earth's surface. When the air reaches the cooler upper layers, its water vapor condenses onto fine particles in the air—such as dust, pollen grains, or pollutants—and forms a cloud.

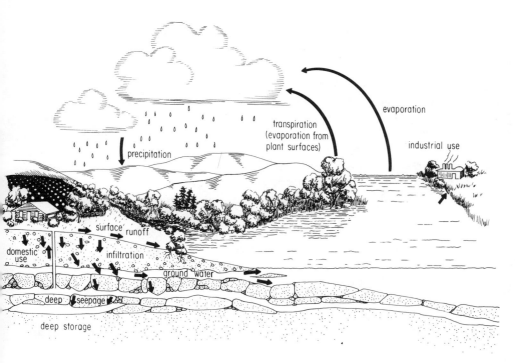

FIGURE 1–7: The water cycle. Arrows show the flow of water through the ecosystem. (Redrawn with permission from Robert Leo Smith, *Ecology and Field Biology*. New York: Harper and Row Publishers, Inc., 1980, Figure 6–2.)

Similarly, when we exhale on a cold day, our warm, moist breath condenses out and a tiny cloud forms for an instant, then it disappears. Clouds do not last forever in the sky, either; new ones keep forming and old ones evaporate. The next time you see cottony cumulus clouds blowing across a blue sky, pick out a tiny, isolated wisp of a cloud and watch it for a minute or two. It will gradually disappear.

If the air is very moist, the droplets of a cloud will grow larger until they can no longer be buoyed by the swirling atmospheric winds. These droplets will fall to earth as precipitation, in the form of rain or snow, sleet or hail, depending on atmospheric conditions. Enough precipitation falls over the earth's surface each year to form a layer of water 3.3 feet (1 meter) deep.[1] More water evaporates from the oceans than falls as precipitation each year, while the reverse is true for land masses. The enormous volume of water that flows through rivers and into the oceans maintains a rough balance between the amount of water contained on land and in the oceans. Even this flow is less than 30 percent of the total precipitation on land; the rest either evaporates once again, is taken up by plant roots, or enters the deeper layers of ground water. Some water may spend years frozen into glaciers or as part of the vast store of deep, still ocean waters.

Water does not remain unchanged throughout its world travels. Evaporation is a natural cleanser, leaving behind most of the substances contained in water on land. As·soon as it enters the atmosphere, however, water begins to mix with other substances. Each raindrop condenses around a minute particle that becomes its center. Being a reactive substance, the weak solutions in clouds can form many compounds, including the naturally occurring carbonic acid. Of more recent concern, since the onset of the industrial revolution, are the stronger nitric and sulfuric acids, or acid rain, formed in rainwater from the nitrogen oxides and sulfur dioxide emitted from burning fossil fuels. Acid rain is discussed more fully in Chapter 2. People do not have a major impact on the water cycle itself, but we do determine the quality of water that circulates globally, sometimes to distant locations thousands of miles away.

On reaching the ground, rainwater demonstrates its true and complete ability to meld with the environment. Water is heated and cooled time and again. It is warmed by the sun, by hot springs, or by the scalding turbines of electrical generating plants. It is cooled by evaporation, by the inflow of an underground spring or mountain stream, or by the chill in a cave.

૨ક*SURFACE RUNOFF*ૐ

As rainwater flows down over the land as *surface runoff*, it always seeks the lowest point. The exact drainage pattern is determined by *topography*, the shapes and patterns formed by local changes in elevation. Where the water follows the lines of bedrock fractures, the streams and rivers run somewhat parallel and can converge at sharp angles to form a *trellis* drainage pattern. The *radial* drainage that occurs on mountains and volcanoes appears from the air like the radiating spokes of a wheel. A treelike, *dendritic* drainage pattern is characteristic of rivers over level and low-lying areas. Some such rivers spread out and form exaggerated meanders called *oxbows*. The area that drains into a certain river system is called a *watershed*. High points of land, such as peaks and ridges, form *drainage divides*, outlining the edges of adjoining watersheds.

Water is not alone on its trek to the sea. It proceeds, sometimes slowly and at other times dramatically, to take the face of the land down with it. Over the millenia the Colorado River has thundered along its course, carrying away unimaginable volumes of earth and etching a geologic history book 1 mile (1.6 kilometers) deep, 200 miles (322 kilometers) long and 4—18

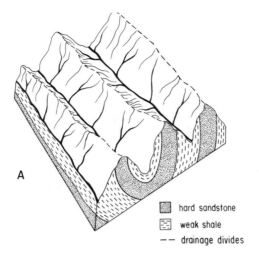

hard sandstone

weak shale

– – drainage divides

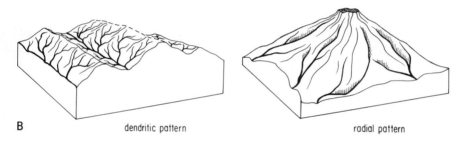

B dendritic pattern

radial pattern

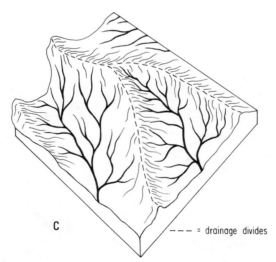

C

– – – = drainage divides

FIGURE 1–8: **A.** Trellis drainage patterns. Cross section of a landscape
showing how valleys have formed where the softer shale has been weathered
more quickly than the harder sandstone, which persists as high points
of land. Drainage divides follow the high elevations between river systems.
B. Drainage patterns. **C.** Aerial view of watersheds showing dendritic drainage
patterns.

miles (6.4–29 kilometers) wide—an area larger than the state of Rhode Island—into the face of the Arizona desert.

Erosion, the transport of soil and rock by water, does not usually occur on such a grandiose scale. Streams and rivers are constantly cutting away at the outside edges of their channels, which often causes the overlying soil to subside into the riverbed and be washed away. Mountain streams tend to cut V-shaped channels with steep sides. Rivers in lower elevations exhibit gradually rounded banks, giving way to broad, flat floodplains on level, lowland areas adjacent to rivers that periodically flood their banks. Human activity, such as farming and construction, contributes greatly to erosion. A fallow field on a moderate slope may look healthy, but can commonly lose tons of topsoil per acre in one year if not protected by the binding action of plant roots.

Runoff can contain suspended particles such as silt: those carried along by its churning action and which will settle out in still water. Dissolved substances are those that become part of the water's chemical makeup and will not settle out. The brown organic acids coming from swamps and bogs are one example of dissolved substances. Nitrogen and phosphorus, two important nutrients present in the soil, are also good examples. Nitrogen enters into solution with runoff, while phosphorus is carried both in solution and chemically attached or adsorbed on soil particles during erosion.

Soil and rock are also changed on site by weathering, which is respon-

FIGURE 1–9: Erosion. A colony of bank swallows have excavated their nest holes into the exposed soil. (*Photo by Michael J. Caduto.*)

sible for soil formation. Weathering is partly a chemical process, during which minerals can be dissolved in solution by the acids contained in rain and running water. Rocks are aged and weakened physically by exposure to sunlight, wind, and rain. Large and small chunks of rock are broken down by water that enters cracks and crevices and then expands upon freezing, exerting enormous fracturing pressures. Wind-driven sand can scour and sculpt soil and rock. By all these forces bedrock is gradually transformed into finer particles and eventually into soil. Through millions of years of erosion and weathering, water and wind have changed the Appalachian Mountains from a range that would have rivaled the Rocky Mountains in height and grandeur into the ancient, rounded, exposed stumps of those mountains that we see today. The eroded sediments from these forgotten mountains are now part of the Atlantic continental shelf.

As it flows through the soil and overland, water adopts the qualities of the minerals that are component parts of the bedrock and soil of that area. Calcium is derived from limestone, magnesium from serpentine and other rocks, and silicon from granitic rocks. Hard water, which carries a lot of dissolved mineral salts, forms over easily eroded, softer rocks such as limestone; while soft water, being relatively free of dissolved mineral salts, forms over hard rocks that are resistant to weathering and erosion. A good way to remember this is "Hard water, soft rocks; soft water, hard rocks." The combined action of plants, animals, and water forms the soil, with its upper layer rich in organic matter and deeper strata composed mostly of rocks and minerals.

≈●GROUND WATER≈●

Beneath the soil surface lies a vast reservoir that comprises 22 percent of the earth's total supply of fresh water, including that which exists as a solid (ice), liquid, and gas (water vapor). The usable ground-water stores in the United States are roughly equal to the amount of precipitation that would fall over that land in ten years.

When water enters the ground it is called *soil water*. Some of it, called *pellicular water*, clings molecularly to soil particles. Capillary action holds still more water between the soil grains, interspersed with minute pockets of air in the *capillary zone*. (See Fig. 1–10, p. 18) The available soil water in this, the *zone of aeration*, is the total amount of water that plants are capable of taking up through their roots. As excess water flows deeper through the porous substrate, that in which pore space is large enough for water to pass through, it eventually reaches an impermeable layer where water cannot penetrate. It then backs up above this layer and fills all the available space, forming a saturated zone. The uppermost edge of this zone is called the

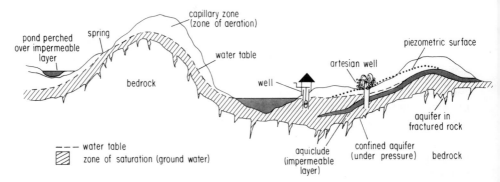

FIGURE 1–10: Relationship of groundwater to land forms.

water table, and all the water within this *zone of saturation* is *ground water.* An *aquifer* usually refers to a reservoir of water contained within permeable material, such as porous rock, sand, or gravel, and through which that water can move freely.

Ground water is constantly flowing downhill, following the contours of the land. Springs form where the water table intersects the surface, discharging ground water to the area above ground. Places where the water enters the ground through porous soils, such as in lakes, rivers, and wetlands, are called *recharge areas.* Recharge is affected by topography and the growing season. When plants are in leaf they intercept large amounts of ground water through their roots before it reaches underground stores. This water is used for plant growth, and much of it is lost due to evaporation from leaf surfaces.

In some places, where the ground water flows down beneath an *aquiclude* or impermeable layer such as bedrock or clay, enormous hydraulic pressures build up. If a well is drilled through the impermeable layer, the water finds an outlet and shoots forcefully to the surface, forming an artesian well. The level to which any given well will rise where ground water is under pressure is called the *piezometric surface.*

Different substrates have various storage capacities for ground water. Some soils, like gravel and sand, have a much greater storage capacity than loam, clay, and other fine-grained soils. Likewise limestone, dolomite, and sandstone hold more water than other kinds of less-permeable bedrock such as metamorphic or igneous rocks. Within the bedrock, water is stored in cracks and dissolved spaces. In limestone areas, where the bedrock is highly soluble, ground water can dissolve away large zones of bedrock, forming underground caves. When such a void is close to the surface, the overlying layers can gradually, sometimes even dramatically subside to form *sinkholes.* An extensive area of sinkholes is called *karst topography.*

FIGURE 1–11: Catastrophic aftermath of the dramatic appearance of a Florida sinkhole. (*Photo courtesy of the United States Department of the Interior Geological Survey.*)

Water is a true artist, working in tandem with the other elements of our world to design a dynamic landscape. As it cuts and dissolves, freezes, squeezes, and carries the rock and soil on its journey, it in turn is altered, taking on the character of the local environment. The physical earth, as we know it, is a giant sculpture on which the infinitely creative powers of water's liquid chisel and penetrating fluids have left their indelible mark.

ᘒ*ADDITIONAL READING*ᘒ

Goldman, Charles R., and Alexander J. Horne. *Limnology*. New York: McGraw Hill Book Co., 1983.

Hay, Edward A., and A. Lee McAlester. *Physical Geology: Principles and Perspectives*. Englewood Cliffs, N. J.: Prentice Hall, Inc., 1984.

Usinger, Robert L., *The Life of Rivers and Streams*. New York: McGraw Hill Book Co., 1967.

❧NOTES❧

1. Edward A. Hay and A. Lee McAlester, *Physical Geology: Principles and Perspectives* (Englewood Cliffs, N.J.: Prentice-Hall, Inc., 1984), p. 306.

FIGURE 1–12: *(Photo by Cecil B. Hoisington.)*

CHAPTER TWO

FRESHWATER ECOLOGY
A Home and Its Inhabitants

R ain had been falling in torrents under bleak, gray skies for two nights and a day. The soft, muffled splashes of raindrops striking oak leaves mixed with the sounds of flowing water as it trickled and dripped its way into the lowlands. By now the shorelines of lakes, ponds, and streams had crept well beyond their normally respected high water marks, flooding roads, marshes, and low-lying homes. Lake Wionkhiege (wee-on-keeg) in Smithfield, Rhode Island, had risen over five feet (1.5 meters) in less than thirty-six hours.

Sometime just before dawn on the second morning of this storm in the spring of 1970, the strained Lake Wionkhiege dam buckled, turning the outflowing stream into a deluge carrying uprooted trees, rocks, and splinters of broken dam along its course. Within several hours the water level had receded to 6 feet (1.8 meters) below its normal height.

This event began a traumatic decade for the wildlife and plants of Lake Wionkhiege. Muskrat homes became exposed high on the muddy banks. The abundant population of fish—yellow perch, golden shiners, largemouth bass, and more—were squeezed into the remaining water, which was a fraction of its original area. Thousands of freshwater invertebrates living in the mud were exposed. Those animals and insects that could do so found their way back to the water or burrowed into the mud; the rest perished. Many emergent aquatic plants, such as arrowhead and buttonbush, were left stranded with dry feet. The common submergent in Lake Wionkhiege, watermilfoil, which was once well below the lake's surface, had become a tangled, exposed mat.

Normally, bird life on the lake had been restricted to king-fishers, a few families of mallards, and an occasional green heron. Yet within several days after the muddy lake bottom became exposed the area was teeming with birds that have seldom, if ever, frequented this lake before: killdeers, great blue herons, herring and ring-billed gulls, black ducks, wood ducks, spotted sandpipers, snowy egrets, and the lesser yellowlegs. The shorebirds were attracted to the expansive shallow pools and exposed mudflats, just as they would normally come into a tidal flat at low tide. In place of marine invertebrates, fish, and plants, they were feeding on their freshwater cousins. The mud was replete with dragonfly and damselfly nymphs, scuds (small crustaceans), tiny pill clams, countless snails, and freshwater mussels, all of which had been uncovered by the sudden drop in water level.

Green herons, kingfishers, and mallards came in greater numbers than had ever been seen before. One morning there were close to 100 mallards crowded into an area measuring roughly 100 by 200 feet (30 by 61 meters)! These birds were gorging themselves on the tender exposed shoots of watermilfoil and other plants that had once been submerged under 8 feet (2.4 meters) of water.

In time the muskrats rebuilt their homes farther down the banks. The exposed flats abounded with flourishing new plant life. Other forms of life gradually adapted to the shallow water and shrunken boundary of the lake.[1]

FIGURE 2–1: Freshwater mud flats form during droughts or seasonal low-water periods and where ponds or lakes are drained. (*Photo by Michael J. Caduto.*)

Building dams is just one of many ways in which we manipulate the fresh water in our surroundings in order to better enjoy the aesthetic and recreational opportunities it provides. In the countryside and in the fresh-water ponds and lakes found in suburbs and city parks, we are clearly a part of the freshwater environment. But what do we know of the aquatic life that is often within walking distance of our homes? To understand the ecology of fresh waters we will look at the life in these dynamic environments in their own homes, keeping in mind that human actions lurk somewhere in the background as a powerful determinant of environmental quality. Later in this chapter, and throughout this book, human activities are studied as an integral part of the world of fresh water.

As you read, remember that ecosystem boundaries are artificially imposed in an attempt to define distinct units that can be more easily studied and understood. Although the next four chapters define and discuss separately ponds, lakes, streams, rivers, and wetlands, these habitats should be looked at as being parts of an intermingled whole. This is especially true of wetlands, which are really integral and vital zones between deep-water habitats and dry land. A pond, for instance, can easily be called a marsh. There is much overlap between these types of aquatic environments.

A naturalist needs to see the differences between types of environments and should be comfortable with the many gray areas in between. The student of terrestrial ecology is encouraged to retain his or her basic ecological understandings and to adapt them to the world of fresh water. This chapter provides a conceptual framework that will help the reader to know what to look for when studying freshwater life. An important first step toward understanding is to develop the eyes with which to see.

❧ BASIC ECOLOGICAL UNDERSTANDINGS ❧

The changes at Lake Wionkhiege temporarily brought about a new community of plants, animals, and birds in the exposed mud flats. In ecological terms, a *community* refers to all the plants and animals interacting with one another and sharing the available resources and restraints within a defined area. A pond could be considered a community, as could a lake, or even the microscopic life that lives in a water-filled footprint in the soft sand along the shore. An *ecological* study of any community is concerned with the interactions that occur between living things and their environment. It is a study of plants and animals in their homes or *habitats*, those places where certian organisms normally live. Indeed, the root of the word "ecology" is *oikos*, a Greek word meaning "home." An *ecosystem* encompasses all the parts of a certain environment, including the living or *biotic* plants and animals, and the nonliving or *abiotic* components, such as soil, water, air, and the sun's

FIGURE 2–2: Ponds are good places to begin studying fresh water because
they are available to most people, small enough to allow thorough
exploration, and full of life! This farm pond was created twenty years ago
by the construction of a dike. (*Photo by Michael J. Caduto.*)

energy. Ecosystems often fade into one another where they meet. It is often
hard to tell exactly where a stream turns into a river. Such in-between
ecological zones are called *ecotones;* they harbor some of the traits of each of
the environments which they join.

Every ecosystem contains *species* of plants and animals, populations that
are capable of mating and producing fertile offspring with other members of
their own kind, but not with members of other genetically dissimilar or
geographically isolated groups of plants or animals. Species of frogs include
bullfrogs, green frogs, wood frogs, and spring peepers. Each species occupies its own ecological *niche* or role that it fulfills in the environment. A
water strider's niche is as a hunter on the pond surface, while a freshwater
mussel filters out microscopic plants, animals, and waste found in and
around the bottom sediments. If a community contains a great number of
species, its diversity is said to be high. Some extreme environments, such as
natural hot springs or polluted waters, support many tolerant individuals of
only a few species, and so the diversity of life they support is small. Environments containing a variety of habitats are better able to provide for the needs
of a diverse community, especially when those habitats are well interspersed.

Any ecosystem, no matter how simple or complex, needs energy to make it go and to make things grow. *Energy flows through* an ecosystem along a complex pathway, being partly used and dissipated as it passes from one living thing to the next. A firsthand look at the journey that energy takes may bring some surprises.

Suppose you are pure energy from the sun. The unit for measuring you is the *photon*. When the sun's rays are striking the earth most directly—which they do during the summer, at midday, and at southerly latitudes—photons will be more concentrated over any given area, bringing more energy.

Let's say that your energy shines down into a pond. Some of you is reflected back into the atmosphere and some is absorbed by the water as heat. Whenever you strike a green plant such as algae, however, something extraordinary happens. Your photons activate a complex chemical reaction in the green chlorophyll pigments of the leaf tissue. It is here that your energy powers a process called *photosynthesis*, the only means on earth that living organic material can be created from sunlight. With the exception of some bacteria, all other living things depend on the organic energy created by photosynthesis in order to live. Oxygen is a vital by-product of photosynthesis. Green plants are called *primary producers* because they are *autotrophs*, capable of manufacturing organic nutrients from inorganic substances.

The formula for photosynthesis is:*

$$12H_2O + 6CO_2 + \text{chlorophyll (acted on by sunlight)} \rightarrow$$
$$6H_2O + 6O_2 + (C_6H_{12}O_6)n.$$

Stated another way,

water + carbon dioxide + chlorophyll (acted on by sunlight)
yields water + oxygen + carbohydrates (starches, sugars, waxes).

Respiration is the process during which organisms metabolize food molecules to get energy for growth and maintenance. Oxygen is consumed and carbon dioxide given off.

Imagine that your photons were enough energy to create 100 pounds of algae in a pond. You might ask, "What is so important about algae, a plant we can barely see unless it is very abundant? Why not cattails or water lilies?"

Algae is as important to a pond, lake, or stream as grass is to a field of

Note: Too much sunlight can cause stress and inhibit photosynthesis.

grazing cows. Most animal life in a pond either eats algae directly or feeds on smaller animals that eat algae. Algae, other green plants, and some autotrophic bacteria, fill the primary niche of *producers*. Animals or plants that eat other plants or animals are called *consumers*. Consumers are *heterotrophs*, getting their energy for *secondary production* from living or dead organic matter that has been produced primarily by green plants. There are *herbivores* that eat plants and *carnivores* that eat the flesh of animals. *Omnivores* are those that eat several kinds of food, possibly including both plants and animals, and they may eat dead organic remains, or *detritus*, as well. The crayfish is omnivorous, eating small fish, insects, and detritus. Scavengers are the omnivores—such as insects, scuds, and mollusks—that act as garbage collectors, consuming dead plants and animals in the bottom sediments. Fungi and other plants that consume dead plants and animals are called *saprophytes*.

But back to the 100 pounds of algae you have produced. As you are eaten by grazing snails, tadpoles, and other herbivores, about 90 percent of your energy is lost to tissue growth, heat production, respiration, and work done by the plant eaters to survive. Only around 10 percent of the energy contained in the 100 pounds of algae is passed on. It will have produced 10 pounds of herbivores! This figure varies from 2–40 percent according to the season, the efficiency of the animal doing the eating, and the caloric content of the plant.[2] This energy loss continues as herbivores are eaten by *first-order* carnivores, those that eat the plant eaters, and the *second-order* feeders, which eat other carnivores. There is less energy available to the carnivores that are higher up on these feeding steps or *trophic levels* in an ecosystem.

A *food chain* can be constructed that shows the various organisms that energy passes through. The second-order and third-order carnivores, such as the trout and kingfisher, have progressively less energy available to them, so fewer of these can be supported by the available energy in an ecosystem. A *food pyramid* can be used to represent this energy loss.

Nevertheless, the food chain and pyramid are oversimplified. Trout eat more than just mayflies, and osprey eat many kinds of fish. A *food web* shows the myriad possibilities for energy flow through an ecosystem. Even more complete are *dynamic food webs*, which show how much food is eaten and the types of food used at each trophic level. Ecosystems with many consumers in the food web tend to be more stable in the face of environmental disruption. For instance, if one type of herbivore, such as the mayfly, was wiped out by pollution, a stream with large numbers of other herbivores would be less affected than one where mayflies were the most abundant plant eaters.

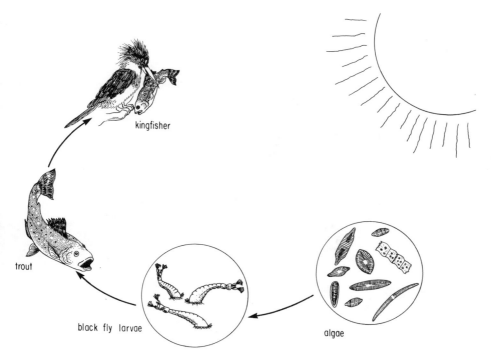

FIGURE 2–3: Stream food chain.

FIGURE 2–4: Stream food pyramid.

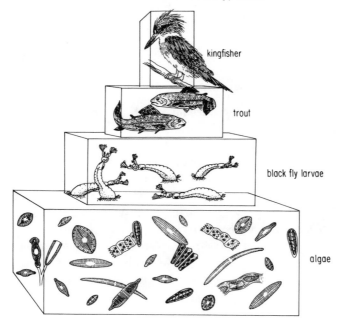

FIGURE 2–5: Stream food web. (1) diatoms; (2) fountain moss (left),
watercress; (3) bacteria and fungi; (4) leaves, sticks, other organic
remains (detritus); (5) snail case caddisfly larva; (6) black fly larva;
(7) zooplankton; (8) snail; (9) mayfly nymph; (10) common sucker;
(11) trout; (12) kingfisher; (13) osprey.

GAS AND NUTRIENT CYCLING

At this point you may be wondering, "O.K., so our energy was consumed as
it scaled the different trophic levels. But what happens to the remains of
those plants and animals that are left over once the energy is gone?"

The inorganic elements that are contained in the environment, and in
the tissues of plants and animals, are essential to supporting life. Enormous
quantities of gases are used in energy production, such as oxygen, nitrogen,
and carbon dioxide, especially during photosynthesis and respiration. *During*
photosynthesis *plants take in carbon dioxide and give off oxygen, while during*
respiration *plants and animals use oxygen and give off carbon dioxide.* The
oxygen/carbon dioxide exchange between plants and animals is an example of
a *gas cycle*—a vital, life-supporting relationship both in aquatic and ter-

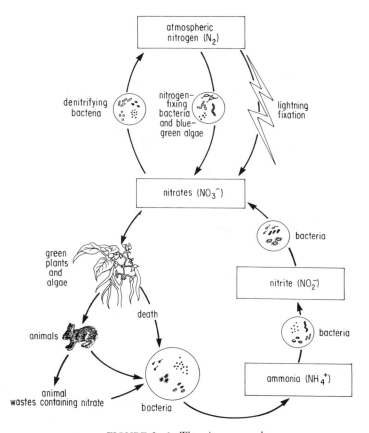

FIGURE 2–6: The nitrogen cycle.

restrial ecosystems. Many elements circulate in the environment via *biogeochemical cycles*.

Nitrogen forms another gas cycle. Five percent of the dry weight of living cells is composed of nitrogen, which is essential for plant and animal growth. Although most nitrogen exists in the form of a gas (N_2), which comprises 79 percent of our air, its most usable forms for plants and animals are ammonia (NH_3) and nitrate (NO_3). Some bacteria and blue-green algae can change nitrogen gas into NH_3 in a process called *nitrogen fixation*. The bacteria *Rhizobium* are associated with the swellings on the roots of legumes and are famous for their nitrogen-fixing capability. Alder and sweet gale also possess symbiotic, nitrogen-fixing root nodules with bacteria from the phylum *Actinomycetes*. Some fungi and the *Pseudomonas* bacteria are *denitrifiers*, changing NO_3 into nitrogen gas and returning nitrogen into the atmo-

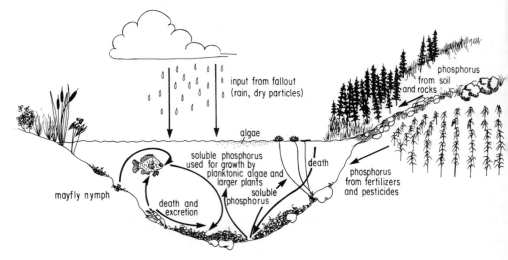

FIGURE 2–7: The phosphorus cycle. The remains of plants, animals, and microbes settle to the bottom sediments. Here they decompose and form an important reservoir of dissolved phosphorus. These nutrients are stirred into the open water during the spring and fall overturns in temperate regions.

sphere. Additional sources of nitrogen in freshwater ecosystems are erosion due to logging, construction, and agriculture; sewage wastes; and fertilizers.

Macronutrients are the elements needed in relatively greater amounts, including phosphorus, calcium, magnesium, potassium, sulfur, chlorine, and sodium. Smaller amounts are needed of other *micronutrients,* such as zinc, manganese, magnesium, silicon, iron, and iodine. These nutrients recycle in the environment.

The *sedimentary cycle* of phosphorus is an example. Phosphorus is a mineral that adheres to soil grains. It can be taken up by plants and is often washed into streams and rivers in eroded soil. Since phosphorus levels are usually low enough to inhibit growth, it is called a *limiting element.* Even a small increase in phosphorus often results in a surge of growth of plants and animals in an ecosystem. This is why the addition of phosphorus into an aquatic environment from erosion or detergents has such a dramatic effect on freshwater life. Luxuriant algal blooms often occur. Some algae can store phosphorus granules and use them during a shortage. Enzymes called *alkaline phosphatases* are used to make phosphorus more available for growth during times of shortage. The constant death of algae and other organisms in lakes and ponds forms a phosphorus reservoir in bottom sediments.

How do the nutrients that are incorporated into plant and animal tissue become reduced to their basic elements in a form once again available for plant growth? This is the job of the *decomposers*. In terrestrial soils, *aerobic* bacteria and fungi—those that need oxygen to live—are the chief decomposers, feeding on detritus. These organisms are in turn eaten by *microbial consumers* such as nematodes, springtails, and mites. Nutrients are released for plant growth through leaching from detritus, via excretion, and upon the death of decomposers in a process called *mineralization*.

Aquatic muds, however, are mostly saturated and devoid of oxygen. The few fungi and bacteria that live here are able to survive these *anaerobic* (without oxygen) conditions. Anaerobic bacteria are more abundant than fungi in aquatic muds, and the incomplete decomposition that occurs here causes organic matter to gradually accumulate. Some anaerobic bacteria are *chemotrophic*, using chemical reactions in order to derive organic nutrients for growth from inorganic elements. The by-products are hydrogen sulfide (H_2S), which smells like rotten eggs, and methane (CH_4) or marsh gas, a colorless, odorless gas that is highly combustible. Nutrients are also released through excretion and elimination by algae, microscopic animals, and the animals that feed on detritus. Their organic remains are also broken down physically and chemically, further releasing nutrients for plant growth. This series of events completes the nutrient cycle and makes continuing life possible.

Although, as a group, the decomposers are poorly known, they form the essential link in the nutrient cycle where organic matter is broken down into its inorganic constituents and made available for plant growth. The reverse of photosynthesis, decomposition closes one of the great and essential circles of life.

❧CONDITIONS FOR LIFE IN FRESH WATER❧

Imagine living underwater. What are the conditions in this home beneath the waves? To solve this puzzle some pieces must first be laid down, creating a picture of what aquatic life would be like.

DISSOLVED OXYGEN

Being the land dwellers that we are, a first thought arises: "Where is the oxygen in water?" Although it cannot usually be seen, water does contain oxygen for those organisms that are adapted to getting it.

Dissolved oxygen (DO) is the amount of oxygen that any given water

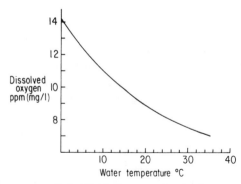

FIGURE 2–8: Relationship between dissolved oxygen and water temperature in a well-mixed system of pure water at sea level and at a barometric pressure of 760mm of mercury (average at sea level). There is less dissolved oxygen at higher altitudes and in very salty water.
(Redrawn with permission from Charles R. Goldman and Alexander J. Horne, *Limnology*. New York: McGraw-Hill Book Co., 1983, p. 103.)

supply or sample contains in solution. It is usually measured in parts per million (ppm), which can also be expressed as mg/l. (One part per million is equivalent to 1 liter placed in a lake that has the surface area of a football field and a depth of 8.8 inches or 22.4 centimeters.) The DO level is affected by many factors. Of major importance is the inverse relationship between dissolved gases, including oxygen and carbon dioxide, and temperature. Warmer water contains less oxygen; conversely, colder water has a higher DO. Some other influences on DO levels include:

- *Photosynthesis and light levels.* During the daytime, sunlight induces high levels of photosynthesis in green plants, especially algae, resulting in the production of significant quantities of oxygen which enter the water in solution. When the sun sets, photosynthesis stops, except for very low levels on brightly moonlit nights. Dissolved oxygen levels are highest in late afternoon and lowest just before sunrise.
- *Respiration.* This process uses oxygen and produces carbon dioxide. Respiration continues throughout the nighttime, so DO levels are usually reduced by morning. Waters with higher levels of organic matter have more respiration occurring during decomposition, which causes lower levels of DO. Waters with rocky bottoms generally have less organic matter and tend toward a higher DO count.
- *Water velocity.* Whenever water is moving, as in streams and rivers, it acts like a blender, mechanically mixing oxygen into the water.
- *Wind.* Wave action and currents caused by wind play a dominant role in churning oxygen into the water; stronger winds enhance this mixing and

generally increase DO levels. Wind is in turn affected by the surface area of a body of water and the surrounding topography. For instance, a large lake with a broad fetch, or open expanse for the wind to blow across, that is situated in flat land is more likely to experience stronger winds than a smaller lake tucked amid hills, which act as barriers that diminish the strength of the wind and decrease atmospheric mixing in these waters.

- *Depth*. This mixing action reaches farther down in shallower waters, increasing the DO, although small whirlpools can carry highly oxygenated water down deep.
- *Ground-water inflow*. Springs bring cold water into a lake or river, but ground water is nearly devoid of oxygen. The immediate effect is to lower the DO near the spring, but, once exposed, the colder ground water can enhance the ability of the surrounding water to hold oxygen.
- *Season*. The seasonal changes in dissolved oxygen levels are complex. They are discussed in the following chapters.

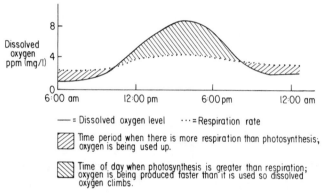

FIGURE 2–9: Daily cycle of photosynthesis and respiration.

Under ideal conditions, in very fast-flowing waters, DO levels can reach a maximum of just under 5 percent of the oxygen that would be held by a similar volume of air.

TEMPERATURE

All plants and animals have their natural limits of tolerance to upper and lower temperatures. There is a lethal temperature that would kill an aquatic organism, stressing its life processes beyond the point that it can adapt to. Over time, however, organisms can slowly adjust to seasonal temperature changes. For instance, a wintering trout would not be able to survive in the warmest temperatures it could tolerate during the summer months; its tolerance to warmer temperatures increases slowly as the stream waters get warmer. The temperature of water is a major determinant of the occurrence of plants and animals. Warmer water increases metabolic rates and respiration rates. Some factors that lower temperature are rapid rates of conduction and convection of heat to colder surroundings, wind evaporation, heavy shade, and the inflow of cold spring water.

pH

Acidity is measured on a pH scale from 1 to 14. This scale measures the strength of hydrogen ions in any solution. A reading of below 7, which is neutral, indicates increasingly greater acidity with decreasing numbers. Above 7 the numbers denote basic conditions. Since this is a logarithmic scale, each increase or decrease of one number means a tenfold increase or decrease in acidity.

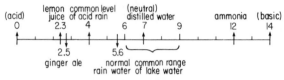

FIGURE 2–10: pH scale.

In aquatic ecosystems, pH has a strong effect on which plants and animals can grow in a certain environment. Most lakes have a pH of from 6 to 9, a range that is tolerable for many plants and animals. However, acid precipitation is increasing acidity to life-threatening levels in some lakes. There are also some naturally occurring conditions, as in bogs, where pH can be very low. Some specialized plants and animals can tolerate high acidity. (See discussion of bogs in Chapter 6.)

Soft-water lakes and rivers, found over granitic bedrock and other rock types with little buffering capacity, are very susceptible to changes in the acidity of waters that feed them. Waters with a high *buffering capacity* are able to resist major shifts in pH levels toward more acid or basic conditions. *Alkalinity* measures the total of all compounds present in water that have the ability to combine with or neutralize acids. Soft rocks, such as limestone, generally increase the alkalinity of the waters associated with them. Many of Florida's lakes are well buffered due to the effects of abundant limestone in that region.

PRODUCTIVITY

Fresh water varies greatly in the amount and composition of nutrients available for plant and animal growth. Productivity is a measure of the fertility or life-sustaining capabilities of waters. This is largely determined by the level of nutrients and light available for plant growth. *Oligotrophic,* a word derived from the Greek *oligos,* meaning small, and *trophe,* or nutrients, is usually used to describe still waters of low productivity. These waters are usually clear, with rocky or sandy bottoms. A *eutrophic* ecosystem is fertile, with abundant nutrients for plant production. Eutrophic waters are high in dissolved minerals such as nitrate, silica, and phosphate, and in suspended matter. *Turbidity* is also high, causing low visibility due to sediments, algae, and other microscopic life that clouds the water. An aquatic environment of moderate productivity is called *mesotrophic.*

Local climates, plant cover, and landforms also affect nutrient levels. For instance, lakes that are fed from watersheds with a larger surface area than other watersheds in a given region are often more productive, because the waters feeding those lakes draw from a greater potential source of nutrients. Climatic extremes of hot and cold can keep soil and bedrock in a disturbed state, which increases erosion and thus the nutrients entering aquatic environments. Temperate climates have a relatively more stable plant cover than arid climates, and less erosion occurs. Plant roots and leaf cover hold the soil in place. Fallen leaves can be a major seasonal source of nutrients in small bodies of water such as ponds and streams.

Hard waters, those containing high levels of dissolved mineral salts, are usually more fertile than soft waters. Some common nutrients present in high levels in hard water are calcium, magnesium, carbonate, and sulfate, as well as nitrates and phosphates.

The ratio between photosynthesis, or production in an ecosystem, and respiration is known as the P/R ratio. When the ratio is greater than 1, an ecosystem has more plant production occurring using the sun's energy than is being consumed by the plants and animals present. A P/R ratio of less than 1 means that consumption is greater than production, indicating that there is an input of organic matter from the surrounding area or from upstream. This input can be natural, as in the fall leaf drop, or of human origin, such as a sewage treatment plant outfall. In this situation, more energy is used up by plants and animals than is produced by green plants from within the ecosystem. The higher levels of respiration produce more carbon dioxide, which reacts with water to form carbonic acid (H_2CO_3), increasing the acidity.

WATER SOURCE AND QUALITY

In comparison to surface runoff, ground water tends to be cooler and cleaner, with few or no bacteria or soil particles present, has little or no dissolved oxygen and is higher in CO_2. It usually is softer, containing less nutrients overall, with those nutrients present more apt to be in dissolved form. The relative amounts of surface runoff and ground water contributing to the total water supply in a stream or lake play a major role in the quality of water present.

❧ FRESHWATER ECOLOGY AND HUMAN INFLUENCE ❧

When people alter an environment, they shift the delicate balance that exists between the biotic and abiotic components. Most forms of development simplify an ecosystem, reducing the diversity of habitats, likewise reducing the abundance of plant and animal species. Food webs become unstrung and rewoven, with fewer energy-flow pathways. The work of decomposers is disturbed, interrupting vital nutrient cycles. In time, the environment may settle into a new equilibrium, but only if it does not incur constant cataclysmic changes, such as the drastic fluctuation of water levels above many dams. Plants and animals, which take long periods of time to adapt to environmental changes, cannot adjust instantly to a constantly changing freshwater environment. Consider the barren, rocky shores of many reservoirs where the water level vacillates several meters or more throughout the year.

Every action we take can affect the quality of our freshwater environments. Although we do not affect the quantity of water that circulates around our fragile planet, we clearly affect the purity of that life-giving element. And since 80 percent of our bodies are composed of fluid, water is truly, as Chief Seattle once forewarned us, "blood which unites one family."

Ecological pollution consists of adding a substance into an ecosystem that is not naturally occurring, or increasing the amount or intensity of a naturally occurring substance in an ecosystem, or altering the level or concentration of a biological or physical component of an ecosystem.

When does a change or addition to a system become pollution? Ecological pollution is present when one of the parameters described above stresses the plants and animals in an ecosystem and requires responses that are beyond their normal range of resiliency. This causes sickness or death and results in a shift in the ecological fabric of that community.

Pollution is a dynamic intruder. The severity of its damage can be affected by environmental conditions such as temperature, water chemistry, season, and the surrounding topography and bedrock. For instance, during the growing season, wetland plants may intercept some pollutants, such as agricultural pesticides or fertilizer runoff, before they reach ponds and streams. When the ground is frozen, plants are dormant and soils impermeable. Pollution runs more freely into the open waterways. Naturally impermeable soils, such as those overlying bedrock, will intercept less pollution as it flows downhill. Hard water tends to be more reactive and can lessen the ecological damage done by pollution.

When several kinds of pollution are introduced into an aquatic ecosystem they can have *synergistic* effects, during which the combined effects can be greater than the sum of the effects of the individual pollutants. A fish that is under stress caused by low oxygen levels in the water is more susceptible to poisoning by insecticides or heavy metals. In some rivers, this combination of pollutants has created zones that are so deadly to fish that they act as walls blocking migration runs. Conversely, *antagonism* can also result when two compounds interact, decreasing the severity of a pollutant's ill effects. Copper, for instance, has been shown to be less toxic in the presence of high levels of calcium such as would be found in hard waters over limestone.

Persistently high levels of pollution through time create communities that consist of many individuals of a few tolerant plants and animals. Some species of *benthic* (bottom-dwelling) invertebrates are often used as indicators of polluted waters. These are not "pollution species," they are organisms that can tolerate conditions such as low levels of oxygen and high sedimentation; lacking competition for available resources, they multiply in great numbers.

There are many ways in which we alter the freshwater environment. A concentrated source of pollution, like the outflow of chemical and industrial wastes from a pipe, is called a *point source*, while general runoff, like that of excess fertilizer from an agricultural field, is called a *nonpoint source*. These are both direct sources of pollution that enter our waterways. Acid rain is an example of indirect pollution, where the activities of people can affect water quality several thousand miles away. Agriculture is responsible for 68 percent of our water pollution, followed closely by human organic wastes and industry.

FIGURE 2–11: (*left*) Nonpoint source of pollution: A municipal garbage dump that is located along the edge of a river. (*right*) Point source of pollution: The discharge of a wastewater treatment plant that flows directly into a river. (*Photos by Michael J. Caduto.*)

NUTRIENT ENRICHMENT

Among the most noxious kinds of pollution are those associated with the addition of inorganic nutrients and organic matter into aquatic ecosystems. Many people have experienced the murky water and offensive odors caused by heavy growth of algae in lake waters that are enriched by the leaching of septic systems from homes along the shore or fertilizer entering from bordering farmland. Phosphate and nitrate are common inorganic nutrients that are often introduced into fresh water via surface runoff that contains agricultural and domestic fertilizers. *Organic pollution* is caused by overburdening an aquatic ecosystem with excess organic matter, such as human sewage waste or manures from stockyards, and from the rich blooms of algae and other organisms that result when inorganic nutrients are introduced in great amounts. *Organic matter* refers to compounds that contain carbon, a basic building block of living things. Nutrients can enter in solution or in suspension, and organic matter can arrive in the form of solids as in leaves, sewage waste, or eroded soil sediments. The major sources of nutrient enrichment are:

- Agriculture, principally runoff and erosion
- Forestry practices: clear cutting and erosion on steep slopes where heavy machines are used, causing an input of mineral nutrients, organic matter, and sediments
- Sewage wastes: effluent from wastewater treatment plants, manure, leaching from septic fields
- Industry
- Pulp and paper mills
- Food-processing plants
- Fires: without protective plant cover, nutrients and soil are more easily washed into waterways
- Urban runoff: the impermeable surfaces of cement and asphalt result in rapid runoff, often introducing lawn and garden fertilizers, pesticides, petroleum products, and other toxics into nearby rivers and lakes
- Natural sources: organic matter entering lakes and ponds from swamps, bogs, and other environments

Healthy ecosystems experience a rough balance between the levels of activity of the oxygen producers and oxygen users. When high levels of organic matter are introduced, the population of aerobic decomposers increases and oxygen is used up faster than it is introduced into the system. Excess nutrients, such as phosphates, nitrates, silicon, and iron, stimulate algal blooms that also reduce DO levels via nighttime respiration and upon death and decomposition of the algae. Thus nutrient enrichment can result in an *oxygen sag*, where anaerobic conditions can occur, stressing fish and other aquatic life. In time, if the pollution does not persist, decomposers will consume the organic matter and the balance between oxygen production and consumption will be restored. Organic pollution is sometimes measured using *biological oxygen demand* (BOD). The BOD of water is an important indicator of how much oxygen-demanding decomposition and respiration will be required to fully consume the organic matter contained in that water.

In still waters where organic pollution is severe, the rates of decomposition are high and DO levels are consistently low. These waters are especially vulnerable during the winter in areas where ice cover decreases atmospheric mixing and photosynthetic production of oxygen. This overfertilization of aquatic ecosystems is called *eutrophication*, and it can result in faster rates of aging in a pond or lake due to high levels of plant production and the buildup of organic remains. (See Ecological Succession, Chapter 3.)

TOXIC ELEMENTS

Toxic elements damage the ability of plants and animals to carry out their life-sustaining functions. The disturbance of organisms short-circuits nutrient cycles and energy flow, sometimes causing disruptions in the workings of an ecosystem. The major sources of toxins are agriculture, industry, paper mills, polluted precipitation, and urban runoff. Pesticides, herbicides, and industrial compounds are the chief contaminants in polluted waters.

Plants and animals have tolerance levels to toxic elements, which, if exceeded, will cause stress, lowered resistance to disease and other environmental hazards, and eventually death. The level of a substance that an organism can tolerate is affected by its initial health, its life history stage, and other environmental conditions such as the season and pollution that may already be present. A trout living in a stream at the upper end of its temperature tolerance range is more susceptible if pesticides enter via runoff from a cornfield upstream. The maximum acceptable tolerance concentration for an organism is determined by long-term studies of individual plants and animals under varying environmental conditions.

Biomagnification is a major ecological concern. This occurs when an element is introduced into an ecosystem and its concentration increases, moving up the food chain. Toxins having a long *biological half-life*, the time that a body takes to rid itself of one half of its load of a substance, are especially dangerous. Long-lived pesticides; such as chlorinated hydrocarbons, become increasingly concentrated in body tissues at higher trophic levels. Clear Lake in California was sprayed with DDD in the 1940s and 1950s to control biting gnats. Plankton absorbed this chemical from the water as they grew and were in turn eaten by filter feeders, which were then consumed by frogs and sunfish. Fish can also absorb toxins through gill membranes. The pesticide became concentrated in sunfish at twelve thousand times the dosage sprayed in the surrounding water. The production of the lake's population of grebes, who fed on these fish, dropped from one thousand actively-nesting grebes in the surrounding marshes in 1950, to no young being produced by 1960.[3] Another chlorinated hydrocarbon, dieldrin, decreases a sculpin's ability to metabolize food.

Evidence shows that concentrations of radioactive substances can also be magnified in food chains. Strontium 90, a radioactive isotope, traveled through the atmosphere as a result of the above-ground testing of nuclear weapons in the 1950s and 1960s. The human body treats strontium 90 in a similar manner to calcium; it becomes concentrated in the bone tissues of those who eat contaminated food, especially milk. Strontium 90 has a biological half-life of one thousand days.[4]

Acid precipitation is an airborne pollutant that has dramatic effects on aquatic life. *Acid rain* is a commonly used catchall term for acid precipitation, which includes rain and snow, sleet, hail, fog, and dry particles that fall from the sky. It is caused by the burning of fossil fuels, especially gasoline and high-sulfur coal and oil. The sulfur dioxide and nitrogen oxides that are produced react with water vapor to form sulfuric and nitric acids. Carbon dioxide, a naturally occurring gas, also reacts with water vapor, resulting in carbonic acid. Normal rainwater has a pH of 5.6; most fish die when lake water reaches a pH of 5 or lower. Since the pH scale is logarithmic, a decrease in one number means a tenfold increase in acidity, and a drop of two numbers indicates rain that is 100 times more acid than normal. Recent measurements in the northeastern United States indicate that rainwater is averaging 40 times more acid than normal, often around a pH of 4.0, although readings do occasionally dip into the upper 3's. Vinegar has a pH of 3.

Is acid rain a new problem? It has been recognized and documented in Europe since the mid-1950s. Air pollution from the heavily industrialized Ruhr Valley, including parts of the Federal Republic of Germany, Belgium, and the Netherlands, is carried on southerly winds into the Scandinavian countries to the north. This pollution forms a potent acid rain. In 1950 a study was begun of 266 Scandinavian lakes, at which time there were 48 that were devoid of fish life. This figure reached 75 by 1960; by 1975, fish were absent in 175 of these lakes. These are mostly soft-water lakes with little buffering capacity.

Acid precipitation turns rainwater into a powerful leaching agent. It can leach nutrients from leaves and can damage the waxy coating on evergreen needles, causing them to lose water and wither. Botanists at the University of Vermont in Burlington have concluded that acid rain is implicated in the death of 50 percent of the spruce trees on Camel's Hump, a picturesque peak in the Green Mountains of Vermont, since 1965. The growth of living spruces has been stunted by one half.[5]

Acid rain can leach copper, aluminum, and other heavy metals out of the soil and into runoff and drinking water. The toxicity of numerous heavy metals has been shown to increase in the presence of acid rain: lead, aluminum, mercury, zinc, cadmium, and copper. Copper concentrations of only 10–40 parts per billion (ppb) can kill fish in acid water. (One part per billion equals the volume of 1 liter placed in a lake that has the surface dimensions of a football field and a depth of 738 feet or 225 meters). Copper interferes with energy metabolism and enzyme function, causes a mucous covering on gills, and can damage kidneys, liver, and spleen. Aluminum is naturally present in the soil and lake bottoms. The increased acidity "mobilizes" this metal into solution, where it binds to fish gills and causes suffocation. Heavy metals can also inhibit the hatching of fish eggs and cause deformed young fish and amphibians to be born, such as frogs and salamanders. If the waters become too acid, algae are killed, eliminating the most important plant food supplying aquatic life. Fish die along with the algae. Most fish do not eat algae directly, but they feed on animals and other small creatures that need algae to live.

Waters over hard rocks, such as granite, are most susceptible to acid rain because there is little buffering capacity. Limestone areas are more resilient. The congressional Office of Technology Assessment has determined that 3,000 lakes and 23,000 miles (37,015 kilometers) of streams and rivers are vulnerable to acid rain in the eastern United States alone.

Acid rain also dissolves the structures of buildings and cars, eroding away iron, steel, limestone, and more.

FIGURE 2–12: Details of the lamb atop this limestone sculpture have been dissolved by acid rain. *(Photo by Michael J. Caduto.)*

Ground-water pollution. There is another kind of pollution that, like acid rain, is often not apparent until its damage is done. Ground-water pollution really describes a place where all the previously described chemical and organic pollutants can go. But because our ground water is such a vital resource, it deserves special mention. About 50 percent of the people in the United States rely on ground-water supplies at home. And for the most part, when we turn on the tap, clear, clean, cool water comes gushing forth: 75 gallons (284 liters) per day for the average user.

Aquifers that are especially prone to contamination are those overlain by permeable soils, through which polluted water can percolate, and those in contact with a water table that is at or near the surface. In addition to the possibility of pollution from those sources already mentioned, ground water can be contaminated by the disposal of organic wastes, such as those from septic tanks and stockyard manure runoff; landfills (dumps) and toxic-waste storage sites from industrial waste; road salt use and storage piles; acid runoff from mines; leaks from underground storage tanks, such as those used for gasoline; and radioactive wastes. Landfills alone can be a source of leachate containing lead, copper, iron, steel, salt, and organic compounds. Ironically, sites that seem to be ideal for waste disposal, such as old gravel pits and natural depressions, are often places where the ground water is at or very close to the surface and is easily contaminated.

PHYSICAL ALTERATION OF FRESHWATER ENVIRONMENTS

Each year more forested wetlands, marshes, and wet meadows are drained for agricultural purposes or filled for construction. Rivers are dredged for navigation. Our rivers and lakes are tapped to provide drinking water and for irrigation, the latter being the largest single use of water in the United States. Dams are built to create reservoirs to supply this water and to generate hydroelectric power. Powerplants that run by heat-generated steam use millions of gallons of water as a coolant each year. Industries use water for production and for dissipating heat that is produced. For instance, one million gallons of water are needed during the production of one thousand barrels of jet fuel.

What are the effects of these environmental alterations? Besides the intended benefits that accrue to human society, the ecological changes wrought tend to cause the death of some ecosystems, a state of constant instability and flux in others, and can sometimes result in the creation of entirely new environments. Consider the common forest management practice that leaves clear-cuts reaching down to, and across, streams and rivers. Eroded sediments during the spring runoff can bury and suffocate the eggs of trout and other spawning fish, and silt can clog fish gills, causing an oxygen-starved death. The increased turbidity of the water decreases sunlight penetration to algae and other green plants, reducing their productivity and resulting oxygen levels from photosynthesis. "What more?" you might ask. When the plants are removed from the banks, and shade is decreased, water temperatures often become too high for trout and many benthic invertebrates, such as stoneflies, to survive.

FIGURE 2–13: Power plants use large volumes of water for cooling processes and return the water to the environment at a higher temperature than is normally found there. (*Photo by Michael J. Caduto.*)

In places where the banks of streams are cleared and the streambed straightened, the natural values of associated wetlands are greatly reduced or eliminated. The cleansing action of the floodplain marshes and swamps, flood-control capabilities, productivity of plants, and fish breeding grounds, are all adversely affected. The uniform bottom of the channel itself offers a lower diversity of habitats for benthic invertebrates, which decreases their productivity and variety. Without the balanced cycle of plant production in riffles and decomposition in pools, productivity is reduced even more. Channelization is an ecological disaster.

Water that is used as a coolant and then returned to the environment at a higher temperature creates *thermal pollution*. This is a serious problem in rivers downstream from the outflow and in ponds and small lakes. The composition of plants and animals in a river can take on the character of communities normally found in warmer water. In small bodies of water in temperate climates, the length of the period of ice cover each year can be shortened. Overall, the metabolic rate of plants and animals increases in warmer water, producing a greater demand for oxygen. Simultaneously, the efficiency of plants producing oxygen, and the amount of oxygen that the water can hold, are both reduced. Overall, there is less oxygen available and a higher biological demand for this limited oxygen.

The many benefits derived from dams—primarily recreation, electrical power generation, and water supply—also have their ecological costs. Many of these drawbacks are subtle. Early dams brought an abrupt halt to the spring runs of salmon, shad, smelts, and other fish that seek the cool, gravel-bottomed streams to spawn. In earlier times, little was understood about fish biology and, even if it had been, technology was not yet available to build large fish-transport mechanisms. In some large rivers, such as the Connecticut River, dams halted the salmon runs in the late 1700s. Salmon tried in vain to spawn for up to a decade after a sixteen-foot dam was built in 1798 at Miller's River, 100 miles up from the ocean along the Connecticut River. Then the salmon disappeared. Prior to this time salmon were so common in some waterways that they were speared during spawning runs and used to fertilize agricultural lands along the banks. Along the Connecticut River today, and in other rivers, the construction of fish ladders and other means of clearing the dams, salmon breeding and reintroduction programs, and pollution controls are slowly bringing back the salmon.

FIGURE 2–14: Fish ladders help migrating fish to reach the headwaters where they spawn. (*Photo by William Azano.*)

The churning water below a dam is often supersaturated with oxygen. This can cause a gas bubble disease in migrating fish, with effects similar to the condition known as the bends, which deep-sea divers experience when they surface too quickly. If water flows over the top of a dam, the temperatures downstream tend to be similar to those normally expected for the river below. Bottom-draining dams, however, especially in very deep lakes where the water comes from near the bottom, cause abnormally cold water downstream, creating localized cold-water communities. Fluctuating water levels, especially marked below flood-control dams, result in lower populations because of the extreme and rapid changes that occur—changes to which few plants or animals can adapt. Tall dams can prevent or inhibit the natural upstream migration of the adult stages of aquatic insects, possibly decreasing populations upstream.

Of course, human alteration of freshwater environments has not been all bad from an ecological perspective. Dams created new habitats for the plants and animals of lakes and associated wetlands. Over 2 million acres (809,389 hectares) of ponds have been created in the twenty-year period ending in mid-1970 alone. With our increasing knowledge of, and appreciation for, the values of freshwater resources, large natural areas have been set aside for their utility to people and wildlife.

❧INTERACTIONS AND ADAPTATIONS❧

Now, with a greater understanding of natural laws and the impact of human activities on freshwater environments, we will look at how plants and animals are adapted for survival in a world of water, and for life with one another. Discovering the infinitely diverse forms of aquatic life brings wonder and mystery, for each living thing is unique in its approach to freshwater life.

On any summer day you can gaze across a pond and see an environment alive with activity. A bass dimples the surface as it catches its meal, an unfortunate insect that has become trapped in the surface tension of the water after falling in from an overhanging branch. Water striders and whirligig beetles race and shoot over the surface, leaving small wakes as they hunt for a meal. The dreamy whir of dragonfly wings startles you as the dragonfly traces an aerial search pattern for food. Beneath the surface are hidden hundreds of plants and animals, giant and small, that form the threads of the pond food web.

In the United States alone over 1200 species of plants can be found that are associated with fresh water. These plants range from microscopic algae to towering cottonwoods. Not to be outdone, the *invertebrate* animals—those without backbones, which make their homes here, finding their larder among the plants and other small animals—number over 10,000 species, and many more are still unidentified. Insects form the main link between the plants, the carnivores, and other animals such as fish. In fact, the average size of all the animal life in the world is about that of a housefly!

Adaptations to freshwater life can be subtle and ingenious. (See Appendix A for a comprehensive list of plant and animal adaptations.) For instance, the blue-green, attached algae growing on the rocks of still waters produce new plants more slowly than those species found in flowing water. In the rapid waters of the stream, algae found on rocks tend to produce new generations quickly to replace those that are swept downstream.

The important role that algae play as the base of the food chain make them an appropriate place to start when considering the many kinds of interactions between organisms. In the intricate weave of life, the food web forms the fabric and energy forms the threads that bind.

Some organisms seem to be found together just by chance, whereas other *symbiotic* relationships involve definite interactions among two or more organisms of different species or type that are living together. In *commensalism*, one partner is helped while the other is unaffected; for instance, a bird using a tree for a nesting platform derives a safe perch for the nest, while the tree is neither helped nor harmed. Another example is the community of water mites, crustaceans, insects, flatworms, and others that thrive in the minute canals of freshwater sponges. When two organisms live together in mutual benefit it is called *mutualism*. An example would be the human intestinal bacteria, *Escherichia coli*, which aid us in digestion while they are in turn nourished by our food. Some protozoans have algae living in their

FIGURE 2–15: A crayfish falls prey to one of its common predators—the little blue heron. (*Photo by Cecil B. Hoisington.*)

FIGURE 2–16: Ecotones are environments where
several kinds of ecosystems join. A diversity
of plants and animals can often be found here.
(*Photo by Michael J. Caduto.*)

gut in a similar mutualistic relationship. *Cooperation* is also used to describe
relationships that are beneficial to those organisms involved. When one
partner is harmed and the other helped, the relationship is normally one of
parasitism or *predation*. A predator kills and eats its prey, while a parasite
slowly feeds on the flesh of the host plant or animal, without killing it.
Predaceous organisms are carnivores.

Competition is another important relationship. Each plant or animal has
specific needs for food, water, shelter, and air, and tolerance levels for such
factors as temperature, light, and current velocity. These needs can be met in
many places, but there is an ideal combination that the organisms will seek.
The overlap between the habitats that fulfill the needs of different organisms
causes them to compete for available resources. This competition is a major
force determining the location of plants and animals in their environment.
Complex communities, those with a great variety of plants and animals, each
trying to fulfill its own niche, tend to be more stable and resilient than those
that are composed of fewer habitats and species. When environmental extremes like drought and food shortage occur, competition intensifies. Overfishing and other human activities can also alter the lines of competition.

Given their environment and interrelationships with other organisms,
plants and animals possess myriad survival adaptations that are readily seen
when visiting the pond, stream, or marsh. Keep your eyes open along the
ecotones between different habitats, where the environment has some of the
character of several different ecosystems. A diversity of plants and animals

meet their needs here. As you read on into the following chapters and visit freshwater environments, you will notice the endless variety of survival adaptations. You might start by asking yourself, "What would I need to live underwater?" Imagine the transformation your body would have to undergo if you were to become a successful freshwater animal, one that could get oxygen, eat, reproduce, move about, and survive seasonal changes—all while living underwater!

The aquatic world is a dynamic and, within limits, resilient environment. Here countless creatures make their homes; the making and breaking of their fortunes is a daily occurrence. The dragonfly nymph that was a predator just moments ago falls prey to a lurking bass. The stream that flowed free and clear when the warblers returned from their wintering grounds has since been dammed by beavers, creating a stillwater world amid the drowned roots of aspen and birch. Human hands, too, are always at work. Recall where we left Lake Wionkhiege earlier in this chapter, and read on about the fate of that lake.

Over ten years after the dam had burst, a new one was built. Slowly, over the winter and early spring, Lake Wionkhiege rose to nearly its former size and depth. The lake's return presented another ecological catastrophe to which the plants and animals had to adapt. Buttonbush, which had declined over the past decade as it turned brown and withered, with fewer green leaves that sprouted each spring, now showed signs of new life. Sunfish and bass returned to their former nesting sites along the shoreline, gently fanning the water to keep silt from settling over the sandy beds of eggs. Leaf boats skittered and whirled across the waters like tiny ships lost at sea. In this way, tranquility returned gradually to the face of the restored Lake Wionkhiege.[6]

ᓚ ADDITIONAL READING ᓚ

Andrews, W. A., et al. *A Guide to the Study of Environmental Pollution.* Englewood Cliffs, N. J.: Prentice-Hall, Inc., 1972.
Boyd, Susan, Lynn Fuller, and Reed Wulsin. *Groundwater: A Community Action Guide.* Washington, D.C.: Concern, Inc., 1984.
Farb, Peter, ed. *Ecology.* New York: Time-Life Books, 1963.
Goldman, Charles R., and Alexander J. Horne. *Limnology.* New York: McGraw Hill Book Co., 1983.

Hynes, H. B. N. *The Biology of Polluted Waters.* Liverpool: Liverpool University Press, 1963.

Hynes, H. B. N. *The Ecology of Running Waters.* Liverpool: Liverpool University Press, 1970.

Irwin, Ronald, and the Members of the Sub-committee on Acid Rain of the Standing Committee on Fisheries and Wildlife. *Still Waters: The Chilling Reality of Acid Rain.* Ottawa, Ontario, Canada: Minister of Supplies and Services, Canada, 1981.

Kormondy, E. I. *Concepts of Ecology.* Englewood Cliffs, N. J.: Prentice-Hall, Inc., 1969.

Odum, E. P. *Ecology.* New York: Holt, Rinehart and Winston, 1963.

Otto, J. H., and A. Towle. *Modern Biology.* New York: Holt, Rinehart and Winston, 1968.

Smith, Robert Leo. *Ecology and Field Biology.* New York: Harper and Row, Publishers, 1974.

Storer, John H. *Man in the Web of Life.* New York: New American Library, 1968.

Vogelmann, Hubert W. "Catastrophe on Camel's Hump," *Natural History* 91, no. 11 (November 1982), 8–14.

≈●NOTES≈●

1. Excerpted from "Low Tide at Lake Wionkhiege" by Michael J. Caduto, *Audobon Society of Rhode Island Report,* 17, no. 4, December 1982–January 1983. Used by permission.

2. Charles R. Goldman and Alexander J. Horne, *Limnology* (New York: McGraw-Hill Book Co., 1983), p. 272.

3. William A. Niering, *The Life of the Marsh* (New York: McGraw-Hill Book Co., 1966), p. 186.

4. Barry Commoner, *The Closing Circle* (New York: Bantam Books, 1972), pp. 47–53.

5. Hubert W. Vogelmann, "Catastrophe on Camel's Hump," *Natural History* 91, no. 11(November 1982), pp. 8–14.

6. Caduto, "Low Tide."

PART TWO
THE STUDY OF STILL WATER ENVIRONMENTS
Lentic Ecology

CHAPTER THREE
THE POND

P onds are places of magic and mystery. They have an aura that draws an inquisitive mind like a magnet. As a child I visited the pond almost daily, spending hours catching frogs, turtles, salamanders, insects—just about anything that moved. Someone on the block once made his own pond at home, in a round swimming pool that was 10 feet (3 meters) wide and 2 feet (.9 meters) deep. This pond had fish, rocks, plants, and turtles; for a month, everything we caught was added to the homemade community. For a while that summer we had captured and transplanted almost every turtle from the natural pond nearby into its backyard counterpart. None of us had ever seen so many turtles, so we set to work painting numbers on their backs to tell them apart. For a year or so after we had released these turtles we could still go to the real pond and recognize each one by number as it was sunning on a rock or log.

FIGURE 3–1: (*Photo by Michael J. Caduto.*)

The pond intrigue need not be relegated to the realm of childhood nature memories. Even a lifetime of pond study could not possibly answer all the questions that may arise. *Limnology* is the branch of aquatic ecology that studies freshwater environments, including flowing, or *lotic*, waters and still, or *lentic*, environments. This body of knowledge has come a long way since the days when it was widely believed that swallows spent the winter hibernating in the pond muds!

¿♣ WHAT IS A POND? ¿♣

Suppose someone asked you to define the word *pond*. You might say, "Well, a pond is a place where the bottom is very muddy and where many different kinds of plants grow, like cattails and water lilies. Lots of things make their home in a pond: Frogs, turtles, dragonflies, snakes, and birds are just a few. And there are many kinds of newts and other creatures hardly bigger than a speck of pepper scurrying around in the water."

This answer touches on the major characteristics of a pond, and it considers a pond to be a distinct ecological unit with certain conditions and boundaries. Although in this chapter the pond will be studied as if it were a well-defined entity, you should keep in mind that there are numerous pondlike environments. Ecologically speaking, a pond environment can exist not only in a well-defined basin, but also in areas that are called marshes; in the still, shallow water on the windward side of a lake; and in the lazy meanders or oxbows of rivers. Many of the same plants and animals are found in these environments. Wetland ecologists refer to these areas as palustrine ecosystems. Because of the prevalence of pond environments and the great diversity of life found in them, in this book more space is devoted to ponds relative to other freshwater environments.

A pond is a shallow body of water with a muddy or silty bottom that generally supports aquatic plant growth from shore to shore. Plants usually ring the shoreline. Because of the shallow water found in a pond, its temperature changes significantly each day during the warm season. A temporary warm layer forms on top in the summertime, which is why a swimmer dives to the bottom to seek cold water. This warmer surface layer often disappears by the time dawn arrives on a cool morning, especially if there is a breeze churning the water. There is also a great variation in dissolved oxygen levels during each daily or *diurnal* cycle, as discussed under photosynthesis and respiration in Chapter 2. The abundance of living things and decomposition occurring in a pond account for its high levels of respiration and carbon dioxide.

FIGURE 3–2: Vernal pond. (*Photo by Michael J. Caduto.*)

A special kind of small pond, called a *vernal pond*, is ephemeral, existing only during the wet spring period and drying up during the summer. Such dramatic changes have caused the organisms that live there to adapt in remarkable ways. One example is the spotted salamander, which comes out during the first few rainy nights of early spring when the temperatures are above freezing. While venturing out on such a night, you will want to take a flashlight and wear warm, dry clothing to search for these cryptic animals. In the glare of oncoming headlights you may also see wood frogs and spring peepers driven by the rhythms of their mating instinct. At times the roadbed seems to have come alive with jumping and crawling amphibians. Male spotted salamanders arrive first and, when the females come, deposit sperm on submerged plants. The adult female takes the sperm into her cloaca, where the eggs are fertilized, then she lays the eggs in the vernal pool, where there are fewer predators to eat the eggs and young than would be present in permanent bodies of water. Their work completed, the salamanders return to a life in the dark world of the soil and leaf litter.

The delicate fairy shrimp, a crustacean, is a sporadic denizen of the

FIGURE 3–3: Spotted salamander (*Ambystoma maculatum*).
Size: 6–7 inches (15.2–17.8 centimeters).

vernal pond. One spring they may be present in great numbers, then they may disappear for many years. A fairy shrimp swims upside down, beating its legs to create a current that moves algae, detritus, and small animals toward its mouth. Eggs are produced for the winter season and for periods of extreme heat and drought. Some of these eggs have hatched after being inactive for over twenty years.

FIGURE 3–4: Fairy shrimp. *Size:* Many species are around 1 inch (2.5 centimeters).

ತಿ THE ORIGINS OF PONDS ತಿ

Some farm ponds were literally born yesterday, while many of glacial origin are the remnants of larger basins that were created thousands of years ago. Here are some of the ways that ponds are formed.

GLACIERS

Many ponds now rest in kettle holes left by the Wisconsin glacier and in the depressions found in ground moraine. Terminal and end moraines some-times act as dams, blocking the flow of rivers and streams to create ponds.

RIVERS

Picture yourself in a canoe on a slowly meandering river through flat ter-rain. Occasionally you approach a bend so sharp that the river almost dou-bles back on itself, then it veers in the opposite direction and continues on its course, forming a looping oxbow. The outer bank of that oxbow is being gradually eroded by the current, which may eventually cut through to the other side of the loop farther downstream, isolating the large bend and forming an *oxbow pond* or lake. Sometimes, as the river channel drifts toward the outer edge of the meanders, sand and silt are deposited on the inside of the turn, forming *scroll ponds* or lakes. Ponds also form where swift rivers enter slower-flowing or still water. Here, the sediment carried by the faster water settles out and builds up to form a *delta*. Deltas form where tributaries enter larger rivers, and at the mouths of rivers where they enter a lake or the ocean.

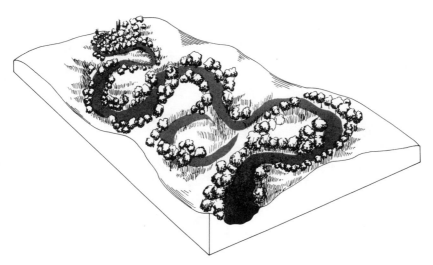

FIGURE 3–5: Formation of an oxbow pond or lake. The
river erodes its outer bank and eventually joins
the channel downstream from the oxbow. A still-water pond
or lake is left in the old channel, depending on the size
and depth of the oxbow.

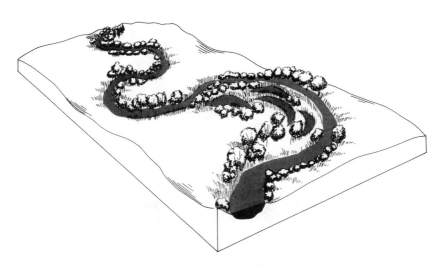

FIGURE 3–6: Formation of scroll ponds or lakes. The
river gradually erodes the outer bank and deposits
sediments on the inside of the bend. Over time the channel
migrates toward the outside of the bend and a scroll
pond or lake forms on the inner edge, depending on the
size and depth of the old channel.

Landslides

Streams that flow through steep-sided valleys are sometimes dammed when heavy rain and spring thaws cause landslides into the valley below.

Animals

Beavers are well known for their ability to create ponds and marshes along stream courses. They will often abandon an old dam when food runs short and build a new one, usually upstream, creating a series of ponds, marshes, and meadows along stream channels. In southern wetlands, American alligators hollow out small ponds during their search for water.

Wind Scouring

Strong winds can scour depressions in the soil that may fill with water. This is especially common around sand dunes.

Ice

In the northern tundra, where permanently frozen ground or *permafrost* is found throughout the year, water cannot penetrate the soil, and even the smallest hollow can form a pond.

People

We are always creating new ponds to provide water supplies for farm animals, lagoons for manure and sewage wastes, and irrigation supplies for field crops. Abandoned gravel pits are another pond source. During a twenty-year period ending in the mid-1970s, there were 2.1 million acres (849,858 hectares) of ponds built in the United States alone.[1]

FIGURE 3–7: A pond that formed in an abandoned gravel pit. (*Photo by Michael J. Caduto.*)

❧THE POND: A FICKLE HOME❧

No two ponds are exactly alike. Living conditions and water quality varies according to the size, depth, and shape of the pond, its persistence throughout the year, the type of surrounding bedrock and soil, local climate and topography, and the effects of human activities.

Conditions in any given pond, too, can change according to the time of day, weather, and season. We have already seen, for instance, that there is a buildup of oxygen during the day, when plants are photosynthesizing, and a decrease at night, when respiring plants and animals use up the oxygen. Wherever cold water, from either a spring or cool stream, is a significant source for a pond, the water remains cooler and is able to hold more dissolved oxygen. In eutrophic waters the daily cycles of oxygen and carbon dioxide levels are more pronounced. There are more plants generating oxygen during the day, which builds up DO levels, and more animals and plants respiring at night, which greatly reduces the DO and builds up carbon dioxide.

Available nutrients are extremely important. Water is a kind of liquid soil; wherever the waters are more fertile, plant and animal growth and abundance are greater. The levels of nitrogen (in the form of nitrate and ammonia) and phosphorus (as phosphate), are the most important determinants of the location of plants and animals. Nitrogen is especially important for the growth of green plant tissue, while phosphorus promotes reproduction—the setting of flowers and seeds. Some micronutrients are crucial for meeting the needs of specific organisms. Iron is important for respiration in both plants and animals, and it is the oxygen-carrying component of hemoglobin in mammalian blood. Calcium is needed for the growth of bones, and silicon is the major element in the shells or *frustules* of diatoms.

Photosynthetic plants and autotrophic bacteria are the primary producers of organic matter in the pond. However, some nutrients and organic matter also enter from outside the pond in the form of leaves and other debris from bordering plants, in stream waters, via insects and migrating animals, and in soil particles that erode into the water. Although nutrients cycle within the pond ecosystem, some nutrient losses do occur through the outflow of the pond and through birds and other animals that feed here and then leave.

The relationship between pond water chemistry and nutrient levels is dynamic. Nutrients tend to be more available for growth at certain pH and DO levels and temperatures. Nutrients, in turn, like calcium and magnesium, can moderate the pH of waters by acting as buffers that neutralize acids. In limestone waters that are thick with plants using up the available carbon dioxide, white calcium carbonate, called *travertine* or *marl*, can precipitate out and cover the plants and substrate.

✃CLASSIFICATION OF LIVING THINGS✃

Before moving on to discuss the life found in ponds, an explanation is needed about how organisms are classified in this book. *Taxonomy* is the branch of biology that is concerned with the systematic classification of living things. Through time there have been numerous systems devised for grouping living things according to such distinctions as cell structure, evolutionary relationships, and details of the reproductive cycles. At one time there were two major groupings or *kingdoms* recognized—*Plantae* and *Animalia*. Then it was noted that bacteria have highly simplified cells that lack a nucleus and other specialized structures, are smaller than other cells, reproduce almost exclusively by asexual division of the cells, and almost always occur as single-celled individuals. These *prokaryotic* cells are distinguished from *eukaryotic* cells, which are larger, contain a nucleus enclosed in a membrane, possess cell structures called *organelles* that have specific functions, reproduce by more complex asexual and sexual means, and are often components of multi-cellular organisms. Bacteria were separated, sometimes along with other organisms, into a third kingdom called the *Monera*. There have since been many other systems devised.

Today it is widely recognized that organisms can be divided into five kingdoms, based on a system that was first proposed by Robert H. Whittaker: *Monera* (bacteria); *Protoctista* (including protozoans and algae); *Fungi* (molds and mushrooms); *Plantae* (flowering plants, conifers, ferns, and mosses); and *Animalia* (animals with and without backbones).[2]

This book uses the terms "plant" and "animal" frequently and, where helpful, gives some specifics to indicate a more detailed classification of organisms. The names of the different divisions used to classify organisms are as follows: moving from the broadest categories to the most specific—kingdom, phylum (botanists use "division"), class, order, family, genus, and species. The three-kingdom system (Monera, Plantae, and Animalia) is the basic classification system used in this book wherever major groupings are necessary, such as in the *List of Common and Latin Names* (Appendix B), and in the organization of each chapter. The reader will find an itemized list of all organisms mentioned in this book, grouped according to the three kingdoms, at the beginning of Appendix B.

Organisms are also discussed in each chapter according to habitat preference and ecological roles. It is hoped that this method of introducing the diverse life of fresh water will be found useful, and that interested readers will explore further into alternative systems used to classify living things.

❧DECOMPOSERS OF THE POND❧

Fungi, and some groups of bacteria, are organisms that share one of the most important ecological niches—that of decomposing dead plant and animal remains into elemental nutrients that will once again be available for plant growth.

BACTERIA

Most scientists include bacteria in the kingdom Monera on the basis of cell structure. Bacteria occur as spheres or *cocci*, rods or *bacilli*, and helices or *spirilla*. These cells divide to form two new daughter cells that are identical to the parent. Since bacteria are among the smallest living things, they usually need to be grown or cultured in colonies before they can be observed. Bacteria can be found in aquatic mud at concentrations as high as one million per cubic centimeter! *Flagella* are used by some species for locomotion; these are hairlike structures that create motion through the water using a two-part beating movement, including a power stroke and a recovery stroke, or some other variation of this method.

Bacteria are more numerous in fertile waters with organic bottoms. As they respire, oxygen is used up and carbon dioxide given off. *Facultative* bacteria can grow with or without oxygen; *obligate* species need oxygen. The reproductive tracts of animals contain *coliform* bacteria, which are common in aquatic environments. Human sewage bacteria, *Escherichia coli*, are indicative of polluted water. *Sphaerotilus* is common in very eutrophic water, while *Leptothrix* thrives where iron levels are high.

FUNGI

These plants grow with long threads called *hyphae*. Fruiting bodies bear spores, which are single cells capable of asexual reproduction. Fungi can also reproduce sexually. Because all fungi lack chlorophyll, they are either saprophytic or parasitic, getting their energy from dead or living plants or animals. Fungi are especially important for their ability to decompose lignin and cellulose, which form the tissues of wood, and chitin, the fingernaillike material that insect exoskeletons are made of. Few bacteria are capable of this important task. Commonly seen on the wounds of fish are the whitish patches of water mold, *Saprolegnia*. (See Fig. 3–8, p. 62) Some fungi emit enzymes to digest their food externally, while others digest internally. *Penicillium* is a well-known aquatic fungus.

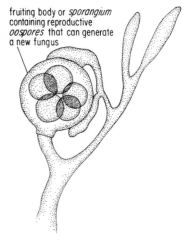

fruiting body or *sporangium*
containing reproductive
oospores that can generate
a new fungus

FIGURE 3–8: A pond fungus (*Saprolegnia ferax*). *Size:* Microscopic detail shown.

❧ POND PLANTS: ALGAE ❧

Take one look at a pond and you see a world dominated by plants. Through photosynthesis, plants use sunlight to create the organic matter that feed pond life; plants also shape the environment, providing homes for animals. Plant cover determines where shade will fall and how much of the water surface will be covered with leaves. They also add oxygen to the water. Individual plants, like the large leaves of water lilies, can support a special community of insects, snails, mites, and other small creatures.

As important as the larger plants or *macrophytes* are in shaping the pond environment, it is algae, the tiniest green plants, that are the chief source of production for the pond. In this section we will look at the algae. Later we will consider the macrophytes that are found in the respective plant zones of the pond. Although similar kinds of plants are discussed together in these sections, they would not necessarily be found growing in the same area.

There are two forms of algae. Some algae are members of the *phytoplankton*, small floating and drifting plants and plant matter, and other forms of algae are attached to rocks, stems, soil, and sometimes even animals. Dragonfly nymphs are often covered with algal growth. Algae lack roots, leaves, and stems. Their main pigments are green chlorophyll, and carotene, which produces orange and red coloration. Some algae are *unicellular*, occurring as individual cells. This is true of many diatoms. Others live in groups or *colonies*, sometimes forming strings called *filaments*.

Euglena propels itself by using whiplike flagella, as do some green algae, while some filamentous algae glide past one another. Diatoms use a stream of protoplasm (cellular fluid) to move. Many algae are capable of both asexual reproduction, during which the parent cell divides to create two new cells, and sexual reproduction, when two cells fuse to form a *zygote*. Certain green algae, for instance, reproduce sexually under stressful environmental conditions to form resting stages that are resistant to drought, heat, and cold. Some algae are active during the winter, though less so than in the spring and summer.

Algae are sensitive to levels of nitrogen and phosphorus in the water. An abundance of these nutrients results in algal blooms, which create great amounts of oxygen during the day. This oxygen is depleted when the algae die and are decomposed by bacteria and fungi, a time when a high level of respiration causes oxygen to be consumed and carbon dioxide to build up.

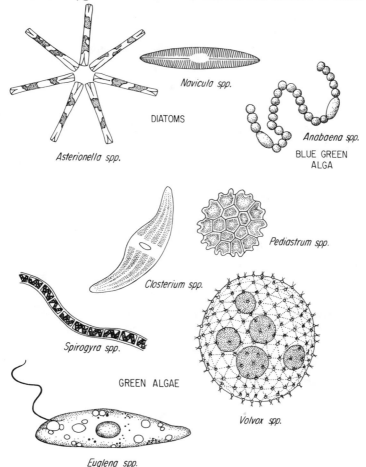

FIGURE 3-9: Some common algae found in ponds. *Size:* Microscopic detail shown.

DIATOMS

If an artist were put to work to design intricate and delicate geometric patterns, diatoms would be the result. Built like a shoebox with a bottom and a lid, diatom shells, called *frustules*, are symmetrical and are composed mostly of silica. These shells are decay-resistant, so diatom remains accumulate in lake sediments. Limnologists study these sediments to discover which diatoms were living at certain previous times. With this information, and knowledge of the habitat preferences of diatoms, a historical picture of the living conditions in a lake can be created.

Diatoms occur in greatest numbers during the spring and early summer. *Asterionella*, with its radial design, is one of the first to appear in springtime. Diatoms are frequently found in moderately fertile ponds. *Fragilaria*, *Tabellaria*, and *Navicula* are common diatoms.

We owe much to these tiny plants. Besides being one of the main plants in the first link of the pond food chain, their remains have provided us with diatomaceous earth for our gardens, and much of our fossil fuel supply was formed from the ancient remains of countless diatoms.

BLUE-GREEN ALGAE

It is now widely recognized that blue-green algae are closely related to bacteria. In fact, they are commonly called blue-green bacteria or *Cyanobacteria* and classified with the Monera. Here they are placed with the other algae because of their important role as photosynthetic producers of the pond.

Blue-green algae have the unfortunate distinction of being known by the sight and smell of their dramatic blooms of mid to late summer, for they thrive in warm, fertile waters. Despite their name, some species turn the water brown or reddish. Some planktonic forms can regulate their buoyancy by adjusting the size of specialized vacuoles that are filled with air. *Vacuoles* are tiny spaces within cells that are surrounded by a membrane and are usually filled with fluid.

Anabaena is a common blue-green alga. Like many blue-green algae, it is capable of nitrogen fixation and can use nitrogen from the air; it can also switch to different forms of nitrogen for its nutrition when this nutrient is scarce. *Microcystis* and *Oscillatoria* are often found, the latter in the shallows along the shoreline. Blue-green algae overwinter as spores.

GREEN ALGAE

This large group contains many diverse and specialized forms. *Desmids,* such as the crescent-shaped *Closterium,* are frequently the predominant algae of oligotrophic waters and are also common in bogs. Another green alga, *Pediastrum,* has a radial design. *Spirogyra* and *Oedogonium* are filamentous. Stonewort (*Nitella*) and muskgrass (*Chara*) are such large green algae that here they are dealt with as submergent plants.

Among the green algae are some tiny plant—animals: They act like microscopic animals but most contain chlorophyll. *Euglena* travels with a flagellum and has a minute eyespot. *Trachelomonas* is vase-shaped; *Phacus* is leaflike, with a red eyespot that is sensitive to light. Most unusual of all is *Volvox,* a spherical colony that consists of up to a thousand cells, each with one eyespot and two flagella. The colony spins as it moves under flagella power.

೩*POND PLANTS: MACROPHYTES*೩

The large and conspicuous plants of the pond are called *macrophytes,* a word meaning "large plants." They consist of two main groups: *bryophytes,* the mosses and liverworts; and *tracheophytes,* plants that have vessels to conduct their fluids. For example, water lilies and cattails are both tracheophytes. *Xylem* provides support for the plant and conducts minerals and water up from the roots, while *phloem* channels carbohydrates, the products of photosynthesis, from the green leaves to the rest of the plant.

Another important difference between these two major groups of plants is their reproductive cycles. In mosses and liverworts the conspicuous generation is the *gametophyte,* that which is capable of reproducing sexually through the fusion of egg (female) and sperm (male). The sperm of bryophytes are motile, and must swim through the water to fertilize the eggs, which then develop into the sporophyte generation. Borne on stalks growing from the sporophyte are *spores,* dust-sized cells that can grow into new gametophytes. In tracheophytes the *sporophyte* generation is most conspicuous—a tree, for instance—and the gametophytes are smaller, such as the egg and sperm (pollen) that form seeds in flowering plants. Many tracheophytes also reproduce with spores, such as ferns, clubmosses, quillworts, and horsetails.

Because of their larger size, macrophytes form distinct habitats in the pond that are convenient to use when learning where and how aquatic plants and animals live. Plants grow where their needs for light, current speed, water clarity, and depth and soil type are met. Interactions with other plants, especially competition, also help determine plant growth. The four major plant zones, moving from the center of the pond to the shoreline, are the open or deep water zone, submergent zone, floating-leaved zone, and emergent zone. These zones can grade into one another gradually, or the transition can be abrupt. Since ponds are not always perfectly circular and sloped, there is usually a mosaic of these zones present, with islands of some zones interspersed in others. This is important to consider, since interspersion creates a greater area of ecotones where species diversity is high.

Following are some common plants found in each of these zones. The open water zone, being dominated by planktonic algae, is not discussed in detail here because it is characteristic of lake habitats.

Emergent Zone and Shoreline

Imagine that you are launching a canoe into a large pond on a clear, sunny day. You have brought along a grappling hook tied onto a long rope for gathering plants. While you are bending down to lift the canoe and carry it toward the water's edge, a strange plant catches your attention. It looks like a miniature, primitive plant that should have become extinct eons ago. Rough ridges run up and down the stem, interrupted by joints that divide the single, unbranched stem into segments. You are looking at the horsetail or "scouring rush," a plant whose ancient relatives lived during the Carboniferous period, the remains of which helped to form our coal deposits. The swollen tips of this, the fruiting branch, will produce spores when mature. Scouring rush stems are so high in silica that they can be used in a pinch to scrub pans and dishes clean.

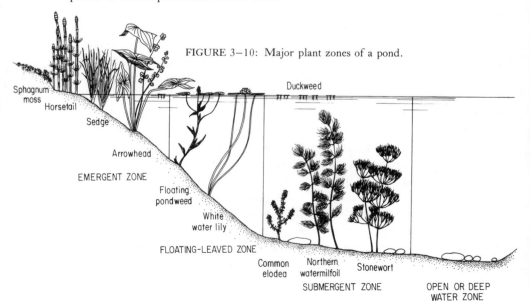

FIGURE 3–10: Major plant zones of a pond.

FIGURE 3–11: Horsetail or scouring rush. *Size:* 11 inches (27.9 centimeters). (*Photo by Michael J. Caduto.*)

FIGURE 3–12: Boat-
leaved *Sphagnum*
(*Sphagnum palustre*).
Size: 2 inches (5.0
centimeters)

FIGURE 3–13: *Sphagnum* moss, showing spore cases. *Size:*
Spore cases measure .1 inch (2.5 millimeters) in diameter.
(Photo by Dr. Cyrus B. McQueen.)

Nearby is some spongy green moss with tiny spore cases held aloft on slender stalks. *Sphagnum* is a common moss in wet areas and is especially important in bogs. Each species of sphagnum has its own requirements for sunlight, water chemistry, and pH.

You may portage your canoe past a low, wet spot that supports a forest of ferns. Some are as tall as 5 feet (1.5 meters). One has a leaf, or frond, that tapers at both ends to form the shape of a giant plume. This is ostrich fern, whose young fiddleheads are a favorite wild edible in the springtime. The base of the stem has a groove in it like a piece of celery; dark, fertile stalks bearing spores arise from the base of the cluster. These stalks will persist through the winter, and are a preferred food for wild turkey. Another tall fern has brown peach fuzz along the base of its stems, and separate fertile stalks as well. This is the well-known cinnamon fern.

Farther along you pass a sunny area covered with a stand of tall grass with ribbonlike leaves and a flowering stalk that towers above the foliage. Careful—the edges of these leaves are quite sharp! This plant, reed canary grass, is common in wet meadows and shoreline areas.

Just before you step off the bank and into the shallow water a fleshy, succulent plant catches your foot. It gets bent over and submerged. You look down to see what it is and brilliant flashes of silver are reflected from the surface of its leaves. It can only be jewelweed, whose leaves are covered with

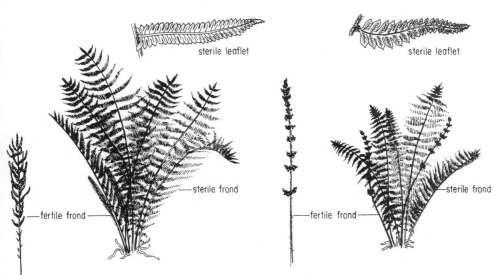

sterile leaflet   sterile leaflet

—sterile frond—   —sterile frond

—fertile frond—   —fertile frond—

FIGURE 3–14: (*left*) Ostrich fern (*Matteuccia struthiopteris*). *Size:* to 5 feet (1.5 meters). (*right*) Cinnamon fern (*Osmunda cinnamomea*). *Size:* to 3 feet (.9 meter).

FIGURE 3–15: Reed canary grass (*Phalaris arundinacea*). *Size:* 2–6.6 feet (.6–2 meters). (*Photo by Michael J. Caduto.*)

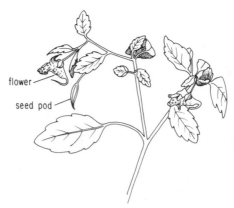

flower
seed pod

FIGURE 3–16: Jewelweed (*Impatiens capensis*).
Size: 2–5 feet (.6–1.5 meters).

a sheet of tiny air bubbles when held underwater, which reflect a silvery sheen from the leaf surface. The flowers of jewelweed, or touch-me-not, form seed capsules that will snap open when touched, throwing the seeds several feet. Jewelweed sap, which is obtained by crushing the fleshy stems and rubbing them on your skin, is an excellent source of relief from the itching caused by poison ivy or the sting of a bee's venom.

Finally you are wending through the shallow water where many kinds

FIGURE 3–17: Emergent plants showing the inflorescence (flowering part) that will later bear seed. (*left*) Tussock sedge (*Carex stricta*). *Size:* 1.6–4.3 feet (.5–1.3 meters). (*center*) Spikerush (*Eleocharis spp.*). (*Size:* varies greatly among species). (*right*) Soft rush (*Juncus effusus*). *Size:* 1.3–6.6 feet (.4–2 meters).

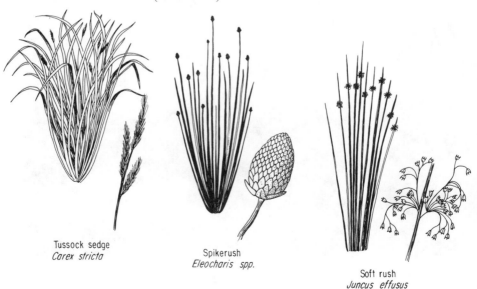

Tussock sedge
Carex stricta

Spikerush
Eleocharis spp.

Soft rush
Juncus effusus

of plants are rooted with their tops emerging into the air. This is the emergent plant zone. Immediately you notice a host of grasslike plants. Tussock sedge forms green tufts that can support a person's weight. The tops die back every year, but the roots overwinter to sprout anew each spring. Spike rush may be growing nearby, with its cluster of long stems capped with a seedhead of hard, three-sided nutlets. Its name belies the fact that it is really a sedge. Soft rush may also be found, bearing a soft inflorescence on the side of the hollow stem just short of the tip.

Confused? You might ask, "How do I tell the difference between a rush and a sedge, and what separates them from grasses?" With a little practice the three can be separated into respective groups using these keys:

- *Grasses:* Have round stems that are usually hollow. The stems are jointed, with leaves attached at the joints by a loose-fitting sheath. They are two-ranked, meaning that if you eye the stem from above, the leaves will appear to be growing opposite one another all the way down the stem.
- *Rushes:* The stems of "rushes are round," often hollow, and their seeds are soft. Leaves are flat.
- *Sedges:* "Sedges have edges" and they are three-ranked. The leaf bases fit the stem snugly and leaves are keeled like a canoe when viewed in cross section. Seeds are tiny, three-sided nutlets.

You notice the sweeping stems of water willow arching from the shoreline. These aggressive plants reach out into open water to form an actively advancing part of the plant community. Their leaves are among the first to change color in the fall, turning a bright red that contrasts with the greens of surrounding leaves.

There are also some shrubs that grow with their roots in the water. Buttonbush, with its dogwoodlike leaves, will bear spherical white blooms in July, comprised of hundreds of small flowers on long stalks. (See Fig. 3–18, p. 72) Later these flowers will form round "buttons." A member of the bayberry family, sweet gale has aromatic leaves like its familiar terrestrial cousin. Conelike seed heads form in large numbers on the stem. Sweet gale is *dioeceous*, bearing male and female flowers on separate plants like a holly. Leatherleaf, *Chamaedaphne calyculata*, grows as a low, compact shrub. Its beautiful Latin name is pronounced "camay-dafney."

The water is nearly up to your knees now, and plant cover is less dense. It is easier to walk, and the canoe is almost ready to glide into open water. Your legs brush by the seedhead of burreed, with its long, thin leaves that rise and fall with each ripple on the water's surface. (See Fig. 3–19, p. 73) Burreed has flower stalks with male flowers forming above the females to facilitate pollination. Cattails, which grow in shallower water, are a close relative of the burreed. Their male and female flowers have a similar ar-

FIGURE 3–18: Buttonbush blooms (*Cephalanthus occidentalis*). *Size:* flowers, about 1 inch (2.5 centimeters) across shrub, 3–10 feet (.9–3 meters) tall. (*Photo by Michael J. Caduto.*)

rangement to those of burreed, with a minor difference between the two species of cattail that can help you to distinguish them: Narrow-leaved cattail has a space between the male and female flowers, while on broad-leaved cattail the flowers of both sexes meet on the stem. Cattails can grow to be 8 feet (2.4 meters) tall; their tubers are a favorite food of the muskrat, while the leaves and stems are used to make its den.

Farther out you find a rise in the pond floor where the water is shallow. A plant with arrow-shaped leaves and a spike of flowers is growing here. This is pickerelweed. Two other plants with arrow-shaped leaves can be confused with pickerelweed, but the flowers and leaf veination can help to separate the three. Arrow arum has an unusual flower and three main veins that radiate from the center of the leaf. Arrowhead or duck potato has white flowers in sets of threes and fanlike veins that run parallel to the leaf edge. Duck potato tubers are a valuable food for waterfowl. These swellings sprout new stems and leaves in the springtime. (See Fig. 3–20, p. 74)

Also in the shallow area a log has lodged in the mud and two small plants have made this their island home. One is a moss, water hypnum, and the other is a leafy liverwort, *Ricciocarpus*. (See Fig. 3–21, p. 74)

FLOATING-LEAVED ZONE

Once you step into the canoe, you hear the sound of water lily pads sliding under the keel. You might see the yellow water lily or spatterdock, with its

flowers
♂
♀

Big burreed
Sparganium eurycarpum

FIGURE 3–19: (*left*) Big burreed
(*Sparganium eurycarpum*). *Size:* 3–6 feet
(.9–1.8 meters). (*below*) Broad-leaved
cattail (*Typha latifolia*), Narrow-leaved
cattail (*Typha angusifolia*).
Size: 6–8 feet (1.8–2.4 meters)

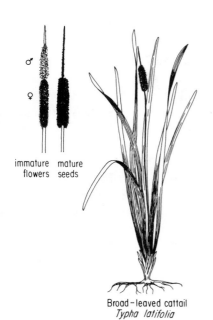

♂
♀

immature   mature
flowers   seeds

Broad-leaved cattail
Typha latifolia

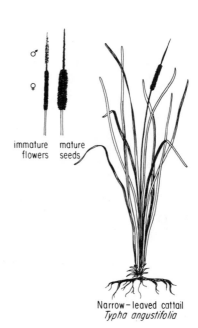

♂
♀

immature   mature
flowers   seeds

Narrow-leaved cattail
Typha angustifolia

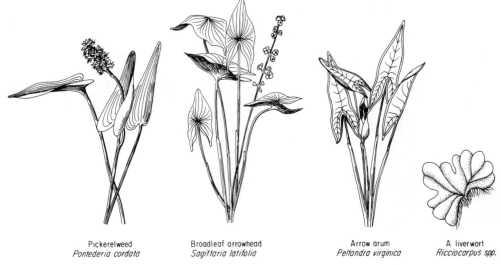

| Pickerelweed | Broadleaf arrowhead | Arrow arum | A liverwort |
| *Pontederia cordata* | *Sagittaria latifolia* | *Peltandra virginica* | *Ricciocarpus spp.* |

FIGURE 3–20: Common emergent plants with arrow-shaped leaves. Pickerelweed (*Pontederia cordata*). *Size:* 1–3 feet (.3–.9 meters). Broadleaf arrowhead (*Sagittaria latifolia*). *Size:* 2–3 feet (.6–.9 meters). Arrow arum (*Peltandra virginica*). *Size:* 2–5 feet (.6–1.5 meters).

FIGURE 3–21: (*far right*) A liverwort (*Ricciocarpus spp.*). *Size:* leaves, .5 inch (1.3 centimeters) across.

oval leaves and yellow flowers that seem to look half closed even when fully open. The large, showy flowers of white water lily draw you toward an expansive stand of these plants, with their floating round pads that have deeply cut, V-shaped notches. Later in the summer, when the water level drops, the flexible white water lily stems will bend and keep the leaves floating on the surface, while the more rigid stems of spatterdock will hold the leaves above the water. The water shield may be nearby, and possibly the American lotus, with its large yellow flowers, 2-foot (.6 meter) leaves and distinctive seedpods. Floating-leaved plants have their pores or *stomata* on the upper surface of the leaf, the opposite arrangement to that of most terrestrial plants. Gas and water exchange occurs through the stomata, and a waxy coating repels water from the leaf surface to keep it dry. Water lily flowers are typically open during the day and have usually closed by late afternoon.

Lily pads are *microhabitats*. The leaves dampen wave action and create calmer water. Many insects make their homes on the undersides of these

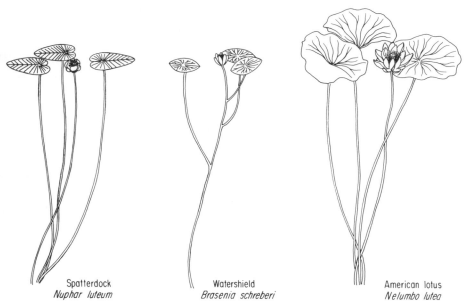

FIGURE 3–22: Water lilies. Spatterdock (*Nuphar luteum*). *Size:* leaves, 6–8 inches (15.2–20.3 centimeters) across. Watershield (*Brasenia schreberi*). *Size:* leaves average 2.5 inches (6.4 centimeters) across. American lotus (*Nelumbo lutea*). *Size:* leaves, 2 feet (.6 meter) across.

FIGURE 3–23: White water lily (*Nymphaea odorata*). *Size:* leaves 7 inches (17.8 centimeters) across. (*Photo by Michael J. Caduto.*)

leaves, as well as snails, water mites, and freshwater sponges. Dragonflies rest here, as do frogs. An aquatic moth larva, the lily leaf caterpillar, cuts out two pieces of leaf and joins them to make a cocoon where it matures into the adult moth.

Tiny floating plants which are among the smallest flowering plants in the world are often found in the still spaces between water lily pads and among the emergents along the shore. Known as duckweeds because of their food value for waterfowl, they are able to get nourishment directly from the water through hanging roots. Even these tiny roots are a microhabitat for algae, water fleas, and microscopic animals. Duckweeds may sometimes cover the entire surface of a pond, forming a footing for insects and spiders. Although the minute flowers enable them to reproduce sexually, duckweeds normally multiply by *budding*, whereby a new plant pinches off from the parent. Great duckweed and lesser duckweed are common, as it watermeal, the smallest of them all. The water fern, which is also called fairy moss, is sometimes found among the duckweeds. This small fern can fix nitrogen with a symbiotic blue-green alga, *Anabaena azollae*.

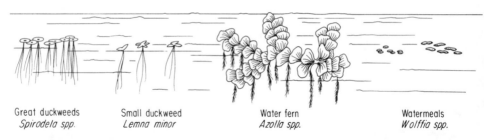

Great duckweeds	Small duckweed	Water fern	Watermeals
Spirodela spp.	*Lemna minor*	*Azolla spp.*	*Wolffia spp.*

FIGURE 3–24: Tiny floating plants that get nourishment directly from the water. Great duckweeds (*Spirodela spp.*). *Size:* .1 inch (2.5 millimeters). Small duckweed (*Lemna minor*). *Size:* .08 inch (2 millimeters). Water fern (*Azolla spp.*) *Size:* leaves, .3 inch (7.6 millimeters) across. Watermeals (*Wolffia spp.*) *Size:* .04 inch (1 millimeter).

Water hyacinth, a relative of our native pickerel weed, is a large, floating plant that was introduced from South America into the waterways of the southern United States. This beautiful and prolific plant, with its flowers of white or blue, forms dense mats over streams and ponds. Although it is food for a few animals, such as the manatee, it casts deep shade over the water below and decreases oxygen production by planktonic algae and submerged macrophytes. By covering the water surface and preventing waves, the atmospheric mixing of oxygen is inhibited as well. The fibrous stems and leaves of the water hyacinth make navigation difficult or impossible in some rivers.

FIGURE 3–25: Water hyacinth (*Eichornia crassipes*).
A widespread floating plant in the waterways of the
southern United States, water hyacinth was introduced
from South America. *Size:* to 1 foot (.3 meter).

Wild celery sometimes grows in the open spaces among the water
lilies. If you throw your grappling hook into the water you can snare some to
have a closer look. The leaves are partitioned into horizontal, fleshy segments. Flowers are found on separate stalks, the male flowers lower than the
females. As the female flowers mature, they grow up to the surface. The

FIGURE 3–26: Wild celery (*Vallisneria americana*). *Size:* leaves to 6.6 feet
(2 meters). The male, pollen-bearing flower breaks loose and floats to the
surface, where the female flower is pollinated. The female flower stalk
then coils up and pulls the flower underwater, where the seeds wil mature.
Floating pondweed (*Potamogeton natans*). *Size:* submerged leaves, 3.9–7.9
inches (10–20 centimeters). Sago pondweed (*Potamogeton pectinatus*).
Size: stems to 3.3 feet (1 meter) long.

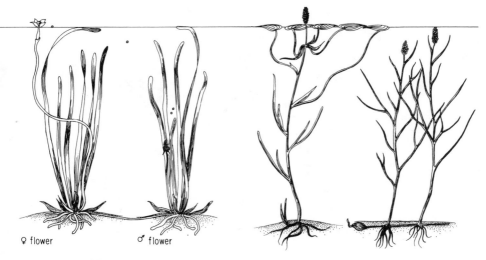

♀ flower   ♂ flower

Wild celery
Vallisneria americana

Floating pondweed
Potamogeton natans

Sago pondweed
Potamogeton pectinatus

male flowers then break loose, float up, and their pollen fertilizes the females. Next, the female flower stalk coils up and pulls the flower under-water, where it will mature into a seedhead.

After studying the fascinating flowering plants among the water lilies, you realize that the open water lies just ahead. A few quick strokes of the paddle and your canoe glides free of the lily pads. Here a plant is growing with two kinds of leaves: small, oval leaves that float on the surface, and narrow ones that are submerged. This is floating pondweed. Sago pondweed grows completely submerged. This plant is an important food for wildlife such as ducks, geese, and coots, which eat its tubers, stems, leaves, and seeds. Only the roots of pondweeds overwinter in temperate regions.

Submerged Zone

You are now floating at the boundary of the floating-leaved and submergent zones. The water is so deep that little sunlight can penetrate and the pressure blow is greater. Wetland ecologists have found that this depth, about 6.6 feet

FIGURE 3–27: Some common submergents.

Common elodea

Common bladderwort

Naiad

Northern watermilfoil   Stonewort

Coontail (hornwort)   Muskgrass   Fanwort   Mermaid weed

(2 meters), is the limit to which emergent plants normally grow. *Emergents* are rooted plants that can tolerate flooded soil but not an extended period of being completely submerged. Beyond here are found submerged plants, and then, if the pond is deep enough, open, unvegetated water.

Your grappling hook hits the water with a splash and soon you are looking at a limp mass of dark green submergents—plants that grow totally underwater. These plants are a true test of your dedication to learning about life in fresh water. At first glance they all look alike, but soon the differences are revealed. The treelike branching pattern of the coontail, the bushy branching tips of common *Elodea,* and the main stem of northern water-milfoil, with fine leaves growing from it, are all good keys for identification. Bladderwort has specialized bladders on its underwater leaves that rest in a deflated position, and then inflate suddenly when a tiny creature touches the hairs around the bladder's opening. The prey is sucked in and then trapped by a flap that closes the bladder, where it is digested. In this zone are found the giant green algae stonewort and muskgrass, the latter usually found in water high in lime.

You should remember that not all of these plants would be found growing together in the same pond, but they would occupy similar plant zones in different ponds.

❧ECOLOGICAL SUCCESSION❧

When your curiosity is satisfied, take a few more strokes out into the center of the pond. Looking back to where you have been, you can see a gradual progression of plant zones from deep to shallow water and finally to the shore. (See Fig. 3–28, p. 80) The deep water you are now floating in will gradually fill in with plant remains and, over time, will become a forested area. The zones you just passed through represent stages of ecological succession. Each plant alters its environmental conditions, such as the soil type, amount of shade cast, temperature, and water level, and makes it more suitable for other species. If you could watch the pond over a long period of time, you would see the submergent zone become shallow enough to support floating-leaved plants and then emergents. A wet, meadow-like environment of grasses, rushes, and sedges would eventually form. Shrubs, and finally trees, would seed in to create a relatively stable *climax community.* The animals found here would also change with the habitat.

Now you can lie back in your canoe and bask in the sunlight.

FIGURE 3–28: A marsh showing several stages of ecological succession: open water, floating-leaved plants, emergents, shrubs, and trees. *(Photo by Michael J. Caduto.)*

Your journey is not over yet. The animal kingdom has thousands of representatives in the pond that are as diverse and challenging as the plants found here. Those animals with backbones, called *vertebrates*, are most conspicuous. Ponds, however, harbor a far greater number of *invertebrates*. Here are the major animal groups.

INVERTEBRATES

Except for mites, eubranchiopods (includes fairy shrimp), insects, and some snails, the ancestors of freshwater invertebrates came from the sea. Numerous invertebrate phyla are represented in fresh water:

- Protozoans or single-celled animals, including the familiar *Amoeba* and *Paramecium*.
- Sponges.
- Hydras and jellyfish.
- Rotifers.
- Bryozoans or moss animals (also classified by some biologists in the phylum *Ectoprocta*).
- Worms and wormlike forms: flatworms, proboscis worms, nematodes, horsehair worms, roundworms, and segmented worms such as aquatic earthworms and leeches.
- Arthropods, the largest animal group, which in numbers comprise 80 percent of all known animals. Included here are the crustaceans, spiders and mites (*arachnids*), and insects. Arthropods have an *exoskeleton* to which muscles are attached, and which must be shed as the animal grows.

  Insects are one of the most conspicuous groups of invertebrates, and so warrant a brief description. An insect's body is divided into a head, thorax, and abdomen. Growing insects, depending upon the species, undergo either *complete metamorphosis*, including the stages of *egg, larva, pupa,* and *adult,* or *incomplete metamorphosis*, skipping the pupal stage and growing from egg to a series of developing *nymphs* (sometimes called *naiads*) to adult. Metamorphosis allows insects to assume different forms and ways of life as they grow. The few aquatic groups that experience complete metamorphosis include the beetles (*Coleoptera*), alder and fishflies (*Megaloptera*), true flies (*Diptera*) and caddisflies (*Trichoptera*). The names of true flies are here written with two words (for example, black fly) and other flylike species as one word (for example, mayfly).

- Water bears or tardigrades live among the plants and sand grains.
- Mollusks include two main groups: the snails (*Gastropoda*) and the clams and mussels (*Pelecypoda*).

Vertebrates

Vertebrates are actually a subgroup of the phylum *Chordata*. Members of this group have a divided nerve cord running along the back or dorsal side, with a segmented backbone, a circulatory system on the front or *ventral* side, and an enclosed brain case. Other characteristics include paired appendages and, in some species, warm-bloodedness or *homeothermy*. Over one half of all vertebrates are bony fishes. Vertebrates also include amphibians, reptiles, birds, and mammals.

❧ POND ANIMALS: DESCRIPTIONS BY HABITAT ❧

Animals generally can be grouped according to where they live. If we know where to look, it is easier to know what to look for. Keep in mind that animals move around according to the time of day, season, and stage of their life cycle. A bullfrog, for instance, spends most of the summer near the shore in the open water at the surface of a pond, then burrows into the mud for the winter. As a tadpole, it frequently rested on the bottom. The three habitats described here are the surface, open water, and bottom.

The Neuston: A Community of the Surface Film

Tiny animals are able to use the surface film of the water as a temporary or permanent home. Many, like the water strider, skitter along the surface, while others hang from it, such as small snails and some species of *Hydra*. The *neuston* refers to all organisms that are found on and under the surface film. For a small creature, the surface tension of the film is a powerful force; microscopic animals may get washed up and stranded on top, where they are unable to break through to reenter the water below.

FIGURE 3–29: Mosquito larva (*Culex spp.*) with breathing tube exposed at the surface. *Size:* .3 inch (7.6 millimeters). Phantom midge larva (*Chaoborus spp.*) *Size:* .4 inch (1 centimeter).

Mosquito larva
Culex spp.

Phantom midge larva
Chaoborus spp.

Mosquito and phantom midge larvae are often found near the surface. Phantom midge larvae are named for their nearly transparent, ghostlike bodies. They prey on crustaceans and other larvae and have flotation sacs to control buoyancy. Adults, who do not eat, can sometimes be seen emerging at the surface from a split in the back of their pupal case. The developing pupae rest on the bottom.

Mosquito larvae, or wrigglers, rest at the surface with their breathing tubes exposed to the air. If disturbed, they will quickly descend by means of whipping motions. Larvae eat algae, microscopic animals, and debris, using mouth brushes to filter their food from the water. An adult female mosquito lays a floating raft of eggs in stagnant waters or puddles. If she has partaken of a blood meal, she will produce an egg mass that is larger, with more viable eggs. Only the female mosquito bites.

Springtails and fish spiders also frequent the water surface. Related to the snowfleas on land, springtails can catapult themselves into the air with a quick flick of the tail. Fish spiders can be occasionally seen scampering among the rocks and plant stems and out over the surface film, buoyed by fine, water-repellent hairs. A large fish spider, at .4 to .7 inches (1 to 1.8 centimeters) across, depresses the surface film dramatically. As it hunts for insects and other small animals, aided by its eight eyes, it can dive for up to forty-five minutes, breathing the air trapped in its thick pile of body hairs. It weaves a web only to raise a brood of up to 300 young.

Several species of true bugs, *Hemiptera*, are members of the neuston. Water treaders are yellow-green and will escape from their shoreline haunts into the open water when disturbed. The female lays eggs inside plant stems. Water measurers step across the water among thick plant cover while

FIGURE 3–30: Springtail. *Size:* .2 inch (5.1 millimeters). Fish spider (*Dolomedes triton*) ♂ *Size:* .4–.7 inch (1–1.8 centimeters).

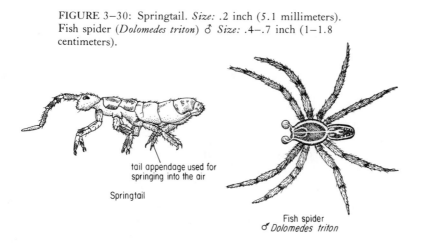

tail appendage used for
springing into the air

Springtail

Fish spider
♂ *Dolomedes triton*

prowling for an insect meal. True bugs have piercing-sucking mouthparts that penetrate a prey's body and allow the bug to suck out the body fluids. Unlike the slow, deliberate movements of the water measurer, water treaders scoot quickly across the surface while they hunt, using the water much as a spider uses its web to catch prey.

Whirligig beetles, members of the beetle order *Coleoptera,* are known for their antics at the pond's edge. Although there are over thirty thousand species of beetle in the United States, whirligigs are the only members of the neuston. These shiny, bluish-black beetles cut a frenetic, gyrating path as they dart over the pond in large groups, their legs beating sixty times each second! Oarlike legs and a unique set of four eyes, two that focus above the water and two that focus below, help them in their incessant search for a tasty inect meal. Occasionally a beetle will dive, using as an aqualung a bubble of air trapped under the abdomen. When the air in the bubble is depleted, more oxygen diffuses in to replenish the supply. The soft aquatic larvae breathe by the diffusion of oxygen through their skin. Resourceful in their survival instinct, some whirligig larvae will even pierce submerged plant tissue in quest of air to breathe. Mature larvae pupate at the pond's edge, producing one generation per year.

air bubble used for
breathing underwater

Water treader
Mesovelia spp.

Whirligig beetle
Dineutes spp.

FIGURE 3–31: Water treader (*Mesovelia spp.*). *Size:* .2 inch (5.1 millimeters). Whirligig beetle (*Dineutes spp.*) *Size:* .5 inch (1.3 centimeters).

Animals of the Open Water

Zooplankton and other microscopic animals. Crustaceans and some species of rotifer are the major components of *zooplankton,* minute animals and bits of animal matter, such as eggs and larvae, that swim and drift in the open water. Many members of the zooplankton are *filter feeders,* animals that eat by sieving fine food particles from the water. Since algae are a staple for these animals, they tend to gather where algae are most abundant. Phytoplankton comprise the first link in the food chain; zooplankton form the second. Zooplankton move deeper during the day, rising to near the surface at night, thus decreasing their chance of being eaten. Temperature, light levels, available oxygen, and water chemistry can affect their location within a body of water. Phantom midge larvae, and fish larvae in the springtime, are periodic members of the zooplankton community.

Several crustaceans abound in the zooplankton. *Crustaceans* are a large and diverse group of animals that each have two pairs of antennae, a pair of mandibles, and two pairs of jawlike maxillae. The structure and number of other body parts vary greatly among the crustaceans. Lobsters, crayfish, aquatic sow bugs, and shrimp are just a few common crustaceans.

One crustacean of the zooplankton, the water flea *Daphnia*, is a member of a group called *cladocerans*. Members of this group swim with five pairs of legs that also move water toward the mouth, filtering out food. Water fleas reproduce quickly and can live for several weeks during the summer, and up to one hundred days if not eaten. They are most common under the ice as winter *resting eggs*. These eggs have been known to hatch after more than twenty years of dormancy! *Parthenogenetic reproduction* is most common among water fleas, a means in which the ovum matures without being fertilized. Many more females are produced than males. Males, however, are needed to fertilize the resting eggs, which are produced during times of environmental stress and in the fall.

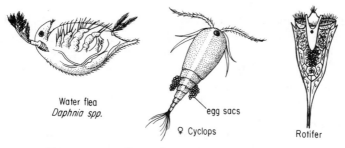

Water flea
Daphnia spp.

egg sacs

♀ Cyclops

Rotifer

FIGURE 3–32: Water flea (*Daphnia sp.*), ♀ Cyclops, Rotifer. *Size for all three:* Microscopic detail shown.

Another crustacean, the *copepod*, is the most abundant winter zooplankter. They grow slowly; a few species produce only one generation each year. The three groups of copepods are the *calanoids, cyclopoids,* and *harpacticoids*. Calanoids are herbivorous copepods, eating algae, and they produce resting eggs for the winter; while cyclopoids, a carnivorous group with biting mouthparts, sometimes produce *cysts*, reproductive structures that can survive harsh conditions and become active when circumstances are favorable. Members of the genus *Cyclops* are well known among the cyclopoids, so named for the single eye they possess in the center of the head. Harpacticoids live in the bottom sediments and vegetation and are uncommon members of the zooplankton.

Rotifers, though not crustaceans, are an omnivorous group with a reproductive cycle similar to that of the cladocerans. Rotifers have one or more rings of cilia that draw food into the mouth. *Cilia* are microscopic, hairlike structures that can aid in movement and are often used by filter feeders to create currents that aid in feeding. When put into motion, the ring of cilia on a rotifer's mouth resembles a turning wheel, thus their name, which comes from the Latin word *rota*, meaning wheel. A few species are predaceous, and many live while attached to plants and even animals.

A rare member of the zooplankton community that bears mentioning is the freshwater jellyfish, *Craspedacusta sowerbyi*. Looking much like its marine relatives, the rhythmic contractions of the bell-shaped medusa move it through the water as it seeks to sting its prey with the 50–100 tentacles along the bell's circumference. This relative of the hydra, and its sessile or attached polyp stage, is seldom found.

Insects. Numerous species of true bugs may be found prowling the pond for a meal. Many of them share the same feeding habits, seizing prey with their front legs and sucking up its body juices. They undergo incomplete metamorphosis. Be careful handling them, because they can produce a sharp, painful sting.

The backswimmer and water boatman are similar in appearance and often mistaken for one another. The water boatman is usually dark gray or black and swims upright, while the backswimmer is generally lighter in color and swims belly up. You may see a backswimmer holding onto a plant stem underwater or floating at the surface, where it gets air through the tip of its abdomen. When submerged, it carries a bubble of air on its abdomen and under its wings. While you watch, it may stroke its oarlike back legs and capture a small insect or polliwog. Water boatmen eat plant remains; they will also suck plant fluids. Their breathing apparatus is similar to the backswimmer's, and is so efficient that they seldomly need to surface, although they must always anchor themselves onto a submerged object to keep from floating up. They are excellent flyers and, if found beneath a street light at night, the male will sometimes chirp by rubbing its legs against its head.

An ingenious snorkel is formed when the water scorpion holds its two posterior appendages together and thrusts them up through the surface. I have seen one of these voracious predators, which are also true bugs, as it ate four dragonfly nymphs in forty-five minutes, leaving skins that looked like shrunken, wrinkled balloons. It is slow in walking, but it ambushes prey with the lightning-quick strike of its forelegs. It, too, is capable of flight. The water measurer looks similar, but lacks the scorpion's breathing tube. Females lay eggs in the stems of rushes.

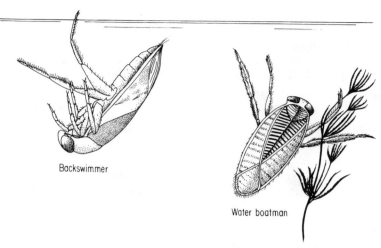

Backswimmer

Water boatman

FIGURE 3–33: Backswimmer. *Size:* .5 inch (1.3 centimeters).
Water boatman. *Size:* .4 inch (1 centimeter)

FIGURE 3–34: Water scorpion. *Size:* 2.5 inches (6.4
centimeters). Water measurer. *Size:* .6 inch
(1.5 centimeters).

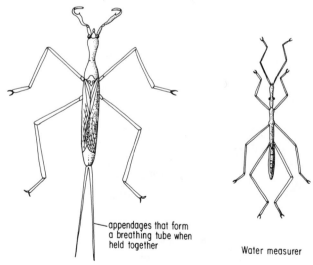

appendages that form
a breathing tube when
held together

Water measurer

Water scorpion

You may be lucky enough to see a giant water bug clinging to the weeds. *Lethocerus* can grow to be 3 inches (7.6 centimeters) long. Another common species, *Belostoma*, only attains a length of about 1 inch (2.5 centimeters). Their brown or dull green, flat, oval bodies are equipped with a venomous, piercing-sucking mouth and a strong instinct to use it. They can be seen either sitting on the bottom, breathing air trapped under their forewings, or resting near the surface with the tip of the abdomen exposed. Large individuals will ambush other insects, tadpoles, crustaceans, or even frogs several times their own size. Nor are they above piercing a bit of exposed human flesh—hence their apt name of "toe biter." Another common name is "electric light bug," after their habit of flying into street lights at night. When tending to its own kind, however, the giant water bug is a gentle beast. Females of the genus *Belostoma* will lay and cement 100 or more eggs onto a male's back. He will carry these eggs for about a week until they hatch. Males have even been known to care briefly for the newly hatched nymphs.

Not to be overlooked, the predaceous diving beetle is raptorial in both its larval and adult stages. It produces one generation per year by laying eggs in plant stems. The larva is long and thin, with a small head. It breathes through its skin and with two openings or *spiracles* on its tail, which it sticks into the air. Pupae are found under stones and plants along the shore. As an adult it uses its third pair of hair-fringed legs as oars. Air is carried under its wings and in a shiny bubble at the tip of its abdomen. Oxygen slowly diffuses into this *physical gill* as it is depleted. Predaceous diving beetles eat tadpoles, small fish, and other insects.

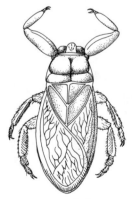

FIGURE 3–35: Giant water bug (*Lethocerus spp.*). *Size:* to 3 inches (7.6 centimeters).

FIGURE 3–36: (*left*) Predaceous diving beetle (*Dytiscus spp.*). *Size:* 1 inch (2.5 centimeters); (*right*) *Dytiscus* larva. *Size:* 1.1 inch (2.8 centimeters).

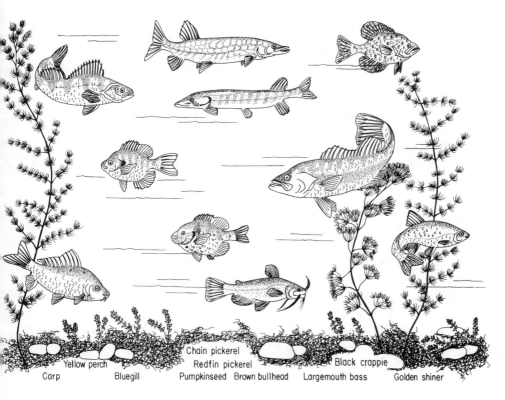

FIGURE 3–37: Some common fish of the pond, marsh, and shallow lakeshore.

Large animals. It is not possible, in this book, to take a close look at all the larger, more conspicuous pond animals. These are the vertebrates, with which people are familiar. Numerous nature-study and field guides are available that give complete life histories of these animals. Some of these books are listed at the end of this chapter. Here is an overview of some large animals that you can expect to see at the pond.

In a sunny spot along the shore a male sunfish, perhaps a bluegill or a pumpkinseed, can be seen fanning the eggs in its nest, a shallow depression hollowed out of the sand and pebbles. Another common sunfish of the weedy shoreline is the black crappie. Sunfish and the largemouth bass, a close relative, will chase off intruding fish or insects as the eggs mature in the warm, shallow water. Bass can grow to be over 2 feet (.6 meters) long and weigh up to 20 pounds (9 kilograms). They will eat frogs and even baby muskrats.

Here, too, is found the carp, a giant member of the minnow family.

Many people are familiar with the rooting habits of carp. These large fish stir up the bottom sediments as they feed, clouding the water and creating conditions under which many other fish cannot live or spawn. Carp were introduced into North America from Europe in 1870.

The brown bullhead, or horned pout, is another nester in the shallows, where the male protects the eggs and newly hatched fry. The bullhead's whiskers, or barbels, are sensitive to touch, helping it to find plant and animal food, both living and dead, along the bottom. Out on the open water you may occasionally see a school of yellow perch feeding at the surface. Perch lay streamers of eggs on vegetation in the shallows.

Chain and redfin pickerel frequent the water lilies and emergent plants. Chain pickerel prefer shallows among the water lilies and sedges, where they feed on frogs, snakes, and sometimes small mammals that enter the water. Their name comes from the dark, chainlike pattern that covers their body, which can weigh from 1 to 4.5 pounds (.5 to 2 kilograms). Their close cousin, the redfin pickerel, rarely grows larger than 10 pounds (4.5 kilograms).

All of these species are bony fish. The most abundant vertebrate group, they have from seven hundred to eight hundred bones in their bodies. Their excellent eyesight enables them to be highly successful predators. Fish have no eyelids, and their eyes are constantly bathed with water. A *lateral line*, located along each side of a fish's body, senses sound and pressure changes.

Frogs, salamanders, and other amphibians are always evident when the summer sun moves across the sky in a vaulted arch. Sunning helps to raise their body temperature to facilitate digestion, growth, and maturation. Sunlight may also help to control external parasites. Amphibians breathe through their skin, which must be kept moist; some have lungs as well.

A bullfrog's bellowing "jug-o'-rum" is familiar to all, a fitting mating call for a creature whose body can grow to be eight inches long and which can live for up to fifteen years. A green frog's call sounds like a loose rubber band being plucked. Other common frogs include the leopard frog, pickerel frog, and the one-inch (2.5 centimeter) spring peeper. Leopard frogs are now rare in some regions because their numbers have been decimated by commercial collectors who supply frogs for classroom study and biological experiments.

As frogs rest on different-colored backgrounds, their skin color changes to similar shades of light and dark. Frog eggs hatch into tadpoles or polliwogs that undergo a dramatic transformation from gill-breathing, algae-grazing young to predaceous, lung-equipped adults. Most tadpoles mature in one year, while bullfrogs take two.

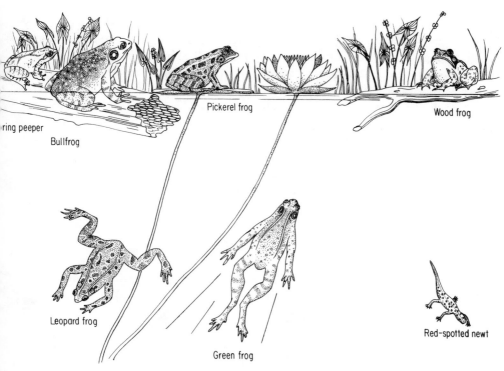

ring peeper

Bullfrog

Pickerel frog

Wood frog

Leopard frog

Green frog

Red-spotted newt

FIGURE 3–38: A sample of many amphibians found in or near the pond.

Although you may not see the land-dwelling adults of the American and Fowler's toads, which only inhabit the pond to mate in the springtime, their tiny, dark tadpoles speckle the shoreline waters during the early summer. Later in the summer hundreds of these minuscule (.33 inch or 8 millimeter) young toads can be seen leaving the pond en masse to begin life on land.

A common predator on amphibian eggs is the red-spotted newt. This is the aquatic adult stage of the immature, land-dwelling red eft which is commonly seen amid the leaf litter and rotting logs of the forest floor. Newt eggs are laid in the shallows on submerged plant stems. Gill-breathing larvae hatch from the eggs. After three to four months have passed, the larvae lose their gills, develop lungs, and take to the land as the brilliant, reddish-orange red efts. Here they live for a few years until they develop a finlike tail, change to a greenish color, and take to the water once again to mate and lay eggs, completing the life cycle.

Turtles often line up on a log to sun themselves, sometimes crawling on top of one another when space is short. Their dark shells are good heat absorbers. Like other reptiles, turtles breathe with lungs. Most reptiles lay eggs with leathery shells and bury them in exposed soil to incubate them, though a few species, like the garter snake, produce live young. Common turtles include the eastern painted, spotted, and mud turtles, as well as the snapper and stinkpot musk turtle. They are mostly scavengers. The plates on their shells grow a new layer each year, while the outside layers slowly wear off.

One of the most conspicuous snakes of the pond is the common water snake, which is quick to anger and will not hesitate to inflict its nonpoisonous bite. The cottonmouth, a poisonous, slow-moving snake of southern waters, is named for the whitish interior of its mouth.

FIGURE 3–39: Reptiles commonly seen around the pond.

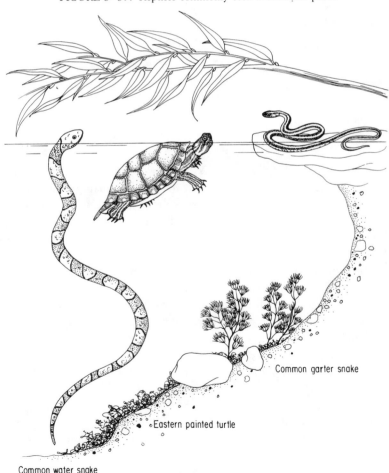

Common garter snake

Eastern painted turtle

Common water snake

FIGURE 3–40: American alligator (*Alligator mississippiensis*). *Size:* Can reach 8–10 feet (2.4–3 meters). (*Photo by Cecil B. Hoisington.*)

American alligators also live in the ponds, marshes, and swamps of the southern United States, and are common in the sloughs along riverbanks. Just the thought of one of these gargantuan reptiles is enough to frighten many people. Yet as ferocious as they appear to be, the females have remarkably strong maternal instincts. They lay eggs and cover them with a nest mound made of wet moss, small sticks, and debris. Nests measure around 5 feet (1.5 meters) in diameter and 2 feet (.6 meters) high. They lay up to forty eggs, each measuring 3.25 inches (8.2 centimeters) long. During the nine-to-ten week incubation period, the female visits the nest; she will even help the young to burrow out of the mound. Females have been known to remain with their young for up to three years.

The abundance of food—succulent water plants, fish, amphibians, and insects—and nesting sites attracts many birds to the pond. Waders frequent the shoreline in search of a meal of fish. Green-backed and great blue herons stalk the shallows, while kingfishers wait patiently on branches overhead. Red-winged blackbirds, purple gallinules, Virginia rails, bitterns, and the energetic marsh wren frequent the cattails and reedlike plants. If you see a shorebird that bobs like a teeter-totter, it is probably the spotted sandpiper, a common sight inland. Shallow, weedy areas provide food for dabbling ducks such as black ducks and mallards, wood ducks and teal. Geese and occasionally swans may be seen here. The comical-looking ruddy duck is often seen among the diving birds, such as common mergansers, ring-necked ducks, coots, and grebes. The kingbirds, a dark flycatcher with a white band on the end of the tail, often hunts insects along with the cedar waxwing, whose velvety feathers and tufted head are unmistakable.

Canada goose
Ruddy duck   Mallard   ♂ Blue-winged teal   ♀ Blue-winged teal

Cedar waxwing   Great blue heron   Red-winged blackbird
                Spotted sandpiper   Eastern kingbird

FIGURE 3–41: Birds of the lakeshore, pond, and marsh.

FIGURE 3–42: The raccoon's masked face is a familiar sight around the pond. *Size* (including tail): 28–38 inches (71.1–96.5 centimeters). (Photo by Cecil B. Hoisington.)

Looking for bird nests takes great skill, but with practice your sense of adventure will be rewarded with sightings of a variety of beautiful eggs and young birds. Birds, having a poor sense of smell, will return to a nest after the eggs are touched, but a human scent may lead a predator to the nest. Because of this, it is better to look but not touch.

Your first thought on seeing raccoon tracks may be that they look like tiny human handprints in the mud. Like most mammals, the raccoon is *nocturnal*, being active mostly at night. The raccoon is a scavenger that will eat fish, frogs, eggs, and crayfish.

A muskrat may slip quietly into the water nearby as it makes for the underwater entrance to its den. It loves to eat succulent submergent and emergent plants and mussels, and is in turn a favorite food for the hawk, otter, and mink. White-tailed deer and moose find food and water at the pond, as do the mink and occasionally playful otter.

The beaver is the most well-known aquatic mammal of them all, and is best seen at dusk, repairing dams, felling trees (a beaver can down a 6-inch [15-centimeter] tree in ten minutes), and seeming never to rest except to eat the inner bark of trees. The beaver has clear eyelids for protection under-water, nostrils with valves to keep out water, and lips that close behind their front teeth to keep the mouth free of wood chips. Webbed hind feet help to propel the beaver, and its naked, scaly tail acts like a rudder. It is kept dry beneath thick, oil-coated fur.

Little brown bat

Water shrew   Beaver

FIGURE 3–43: The abundant life and water found in the pond and
the protection afforded by its waters attract mammals of
the land and sky, as well as aquatic mammals like the beaver.

Active at dusk, *crepuscular* little brown bats hunt over the pond for
insects. Below, a water shrew dives into the pond in search of insects and
small fish, while a star-nosed mole shovels through the wet soil on shore.

All pond animals must adapt to the extremes of winter—a time of cold
weather and food scarcities. Food energy is hard to find and quickly used up
during the hunt and in keeping warm. Many birds escape winter's hardships
by migrating south, while some warm-blooded animals, or *endotherms*, are
capable of remaining active all winter long—for instance, certain birds and
some mammals, such as the beaver, mink, otter, deer, moose, and muskrat.
Mink continue to hunt the shoreline; under the ice, muskrats emerge from
their lodges and forage on plant roots and stems. Underwater stores of stems
were instinctively stockpiled by the beavers during warmer weather. If their
food supply holds up, beavers need only travel to and from the relatively safe
underwater entrance of their lodges to gather food from their cold-weather
caches.

For other animals, *hibernation* is the key to winter survival. While it is hibernating, an animal's metabolism and rate of growth decreases markedly, including heartbeat, respiratory rate, circulation, and body temperature. Little brown bats look for damp, relatively warm and windless places in which to winter over.

Most cold-blooded animals, called *ectotherms*, have to resort to hibernation to live through the winter. Many amphibians and reptiles—frogs, turtles, and salamanders—burrow into the mud of the pond bottom to spend the winter. A few will overwinter beneath the leaf litter or in an abandoned burrow in the forest. Green frogs have been found hibernating both on land and in pond mud. Occasionally a number of frogs, salamanders, and snakes will share the same burrow for the winter. Garter snakes will sometimes hibernate submerged in empty crayfish burrows at the edge of a pond. Some animals, especially painted turtles, can be seen under the ice as they scurry about doing their curious winter errands.

BOTTOM-DWELLING ANIMALS

Among the plants on the bottom of the pond is a rich fauna waiting to be discovered. The *benthic* or bottom environment of the shallows is warmer, yet has more oxygen than deeper waters because of the mixing action of waves and the oxygen-producing plants found there. Nevertheless, planktonic green and blue-green algae occasionally grow so thick in late summer that little sunlight reaches the bottom. Food, especially in the form of detritus, is abundant. Detritus is important in the diet of most benthos or bottom-dwelling animals. These animals change dead organic remains into food for their larger predators, and their wastes help to recycle nutrients back to the plants. Benthic animals are most numerous in the shallows, among the large plants which provide surfaces for attachment, food of attached algae, and good places to hide from predators.

Protozoans consist of a single cell. They are most common in fertile water. Some contain commensal algae and are considered plantlike, such as a *Paramecium* with the green alga *Chlorella* (sometimes called *Zoochlorella*) in its digestive cavity. Most have a red eyespot and are attracted to light. *Paramecia* move by means of numerous hairlike cilia that wave in unison. Others move by means of whipping flagella, while the *Amoeba* extends its *pseudopodia* (false feet) in a process called *streaming*. This is also how the *Amoeba* surrounds its food, enveloping it in a food vacuole in which digestion occurs. *Suctoria* are carnivorous, using their tentacles to trap and eat other protozoa. During harsh environmental conditions, protozoans can form a cyst with a wall that is resistant ot drying and freezing. These cysts

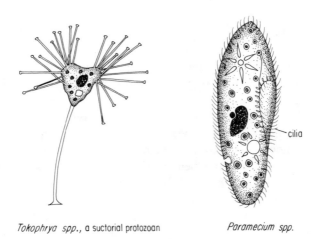

Tokophrya spp., a suctorial protozoan   Paramecium spp.

An amoeba streaming

pseudopodia   food vacuole

FIGURE 3–44: Protozoans. *Size:* Microscopic detail shown.

are thought to travel on the wind and via large animals, which accounts for the worldwide occurrence of many protozoans.

Microscopic water bears or *tardigrades* are sometimes found along the shore in the moist foliage of mosses and liverworts. So named for their lumbering, bearlike gait and general body shape, there the resemblance ends. Their eight legs each have two claws on the end which help them to move as they forage on plants. Two eyespots are found on the head end. Males and females mate to produce fertile eggs which can stand long periods of drying. Adults, too, can remain dried out and dormant for a time, only to become active again once water is available.

While wading in the shallows of an unpolluted pond you may grab the end of a submerged, waterlogged stick to see what is growing on it and find that it is very heavy. A ballooning colony of moss animals, *bryozoans*, is attached that has grown to be 12 inches (30.5 centimeters) across. Starting from a single drought- and cold-resistant *statoblast* that was produced the previous fall, the members of the colony have divided continually to form a mass several thousand strong. Tentacles and beating cilia move food toward their mouths. *Pectinatella* is a common type of moss animal.

Most of the larger animals can be caught using an adept eye, a nimble hand, and an ordinary large metal food strainer. Grab your strainer and scoop out some mud and plant remains. Now the fun begins. Place the mud in a white pan or bucket and begin gently separating out the muck and plants from the animals.

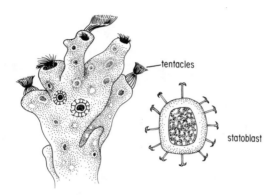

FIGURE 3–45: A bryozoan, *Pectinatella magnifica. Size* of each zooid (individual): .2 inch (5.1 millimeters). The statoblast is produced in early fall; it overwinters to form a new colony of bryozoans in the springtime. *Size* of statoblast: .04 inch (1 millimeter).

Several wormlike creatures squiggle down and out of the light. Identifying these can be confusing, for some are worms, others the larval forms of insects. Some of the major ones to look for can be seen in Figure 3–46.

Two insect larvae that look like worms are the rat-tailed maggot—the larva of an insect known as the hover, flower, or syrphid fly—and the bloodworm, larva of a species of true midge. Rat-tailed maggots get their name from a three-pieced, telescopic breathing tube attached to the posterior end. This tube can be extended to four times their body length. These larvae eat debris. Bloodworms have hemoglobin in their blood, imparting a red color. They eat vegetation and a few live in a tube made of plant debris and sediment. Both of these larvae do well in low-oxygen waters.

The blood of the *Tubifex* or sludgeworm also contains hemoglobin, increasing its oxygen-carrying efficiency and making it well suited for oxygen-depleted or polluted water. *Tubifex* worms live in a tube, head down, eating muck.

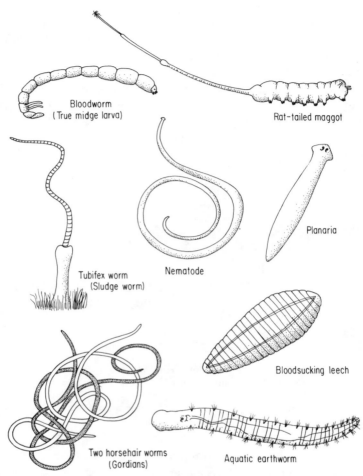

FIGURE 3–46: Worms and wormlike organisms. Bloodworm (true midge larva).
Size: .3 inch (7.6 millimeters). Rat-tailed maggot. *Size:* 1.5 inches
(3.8 centimeters). *Tubifex* worm (sludge worm). *Size:* 1 inch (2.5 centimeters).
Nematode. *Size:* .1 inch (2.5 millimeters) and smaller. Planaria. *Size:* .4
inch (1 centimeter). Bloodsucking leech. *Size:* Various leeches can range
from 3–10 inches (7.6–25.4 centimeters). Two horsehair worms (Gordians).
Size: 4–8 inches (10.1–20.3 centimeters). Aquatic earthworm. *Size:* .5 inch
(1.3 centimeters).

With a little practice you may also find some of the other worms in your sample of pond bottom. Tiniest of all are the abundant *nematodes* or roundworms, which move through the mud eating plants and animals. A small eyedropper will be helpful in separating them from the sample. Many people are familiar with flatworms or *Planaria*. These curious worms have a spade-shaped head with eyespots and a combination mouth/anus about half-way down the ventral side. It uses this mouth to suck the fluids of tiny living and dead animals. Colored like the mud, they are well hidden as they move with cilia on a mucuous coating, or by muscular contraction. One species of *Planaria* can eat *Hydra* and then incorporate the *Hydra's* living, stinging tentacles into the surface of its own body.

If you notice a long (4–8 inches; 10–20 centimeters), stiff worm, it is probably the horsehair or Gordian worm. Masses of horsehair worms will writhe together in a twisted "Gordian knot." Since they do not eat as adults, they have a small or nonexistent mouth. The male, which is curved at one end, mates with a female; eggs are laid on vegetation. When an egg is eaten by a grasshopper or cricket, it hatches and the young larva becomes attached to the insect's gut, where it feeds until ready to emerge an an adult.

Earthworms, too, have their aquatic counterparts that look and live like their terrestrial cousins, eating soil and producing egg-bearing cocoons. Another relative of earthworms, leeches, may also be present. Of the forty-four species found in the United States, most are predators or scavengers. They have a front and a rear sucker and can move inchworm-style or by swimming. Bloodsucking species make a small incision and use a painkiller and anticoagulant called *hirudin* on the unsuspecting host. Once used medicinally for bloodletting in Europe, they were raised by the thousands in leech ponds. They can travel on the feet of birds and move through waterways.

Keep looking and you are bound to see some crustaceans in your mud, maybe even a tiny crayfish. (See Fig. 3–47, p. 102) Like most crustaceans, crayfish are omnivorous, using their two front claws for getting food. Crayfish are *decapods* ("ten feet"), having ten pairs of legs which will regenerate if lost. This occurs gradually during each molt of the exoskeleton. Their cast-off skins are commonly mistaken for dead crayfish. Gills are found on the underside near where the female carries her clutch of up to 700 eggs, which are fertilized externally. She will carry the young crayfish for a time, most of which will only live for about one year. Although they are food for many larger animals, including the raccoon, otter, and numerous fish, they are hard to catch. Their stalked eyes can spot danger coming and they flee backward with a deft flick of their powerful tail. The two pairs of antennae can detect odors.

Scuds or sideswimmers (*amphipods*) are small crustaceans of .2–.8 inches (5–20mm). These creatures are omnivorous; they can swim on their

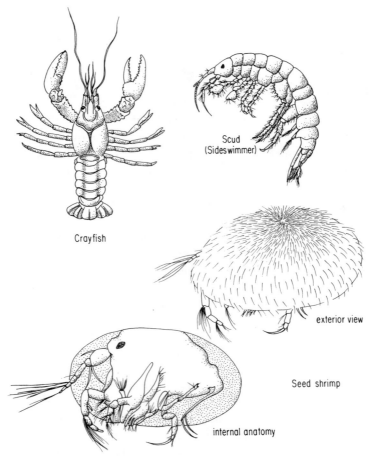

Crayfish

Scud
(Sideswimmer)

exterior view

Seed shrimp

internal anatomy

FIGURE 3–47: Crustaceans found on or near the pond bottom.
Crayfish. *Size:* 4–5 inches (10.1–12.7 centimeters). Scud
(Sideswimmer). *Size:* .2–.8 inch (5.1–20.3 millimeters). Seed
shrimp. *Size:* Microscopic detail shown. (Seed shrimp redrawn with
permission from Robert W. Pennak, *Freshwater Invertebrates
of the United States.* New York: John Wiley and Sons, Inc. 1978,
p. 411.)

sides through the mud and bottom vegetation. Seed shrimps (*ostracods*) are
also abundant in the bottom sediments and vegetation. Although they look
more like miniature clams, a closer look reveals that there are legs sticking
out of their shells. But their internal structure (Figure 3-47) really tells why
these are crustaceans. Seed shrimps are occasionally so numerous that they
form a cloudiness in the water just above the bottom.

Insects, too, are abundant in the benthos. The predaceous whirligig
beetle larva, which will live on the surface as an adult, crawls along the

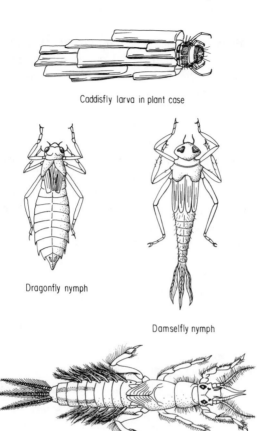

Caddisfly larva in plant case

Dragonfly nymph

Damselfly nymph

A pond mayfly nymph
Ephemera varia

FIGURE 3–48: The pond mud is home for a diverse insect fauna. Caddisfly larva in plant case. *Size:* 1 inch (2.5 centimeters). Dragonfly nymph. *Size:* 1.4 inch (3.6 centimeters). Damselfly nymph. *Size:* .8 inch (2.0 centimeters). A pond mayfly nymph (*Ephemera varia*). *Size:* 1 inch (2.5 centimeters).

bottom plants. If you see what looks like a small pile of stems or leaf pieces walking by, pick it up and the larva of a caddisfly will be peering out of a hole on one end. These omnivorous insect larvae build their own homes out of silk, saliva, and plant parts. Nearby, a mayfly nymph may be munching on leaves or grazing algae from plant stems. Mayfly nymphs have gills on their abdomen; on many species these gills beat constantly. Both caddisflies and mayflies are also common in streams, and are described in more detail in Chapter 5.

Nymphs of the dragonflies and damselflies, the *odonates,* cannot be overlooked. As adults, these are the familiar "darning needles" that children are often warned will sew their mouths shut if they do not keep quiet. They have descended from among the earliest known ancestors of insects, called *Meganeura,* which lived 350 million years ago during the Carboniferous period and had a wingspan of 30 inches (.8 meters). The adults can be seen flying in tandem or resting on plant stems as they mate. The females of some species lay their eggs in plant stems, while others fly low over the water and dip their abdomens in to deposit eggs, which sink to the bottom. Damselfly nymphs, *zygopterans,* have three tapelike gills on their posterior and, like the dragonfly nymphs, they hunt small tadpoles such as those of the American toad, other insects, and crustaceans, using their long and quick-striking *labium,* which grabs the prey. The labium of damselfly and dragonfly nymphs is a long, hinged mouthpart that shoots out, pierces the prey with two pincers, and then draws the unfortunate meal in to be eaten. After a year or two, depending on the species, the nymphs crawl up onto a reed, shed their skins, and emerge as adults. Damselfly adults are slender and hold their wings together over their backs as they wait quietly to ambush other flying insects. But dragonfly adults, *anisopterans,* are the true aerial masters of the insect world. Many hunt on the wing, catching their prey in a basket formed by their legs. Each pair of wings, front and back, is held open when the dragonfly is at rest, and they flap alternately in flight, enabling them to hover and even to fly backward. Few insects can escape their excellent vision with compound eyes of up to 30,000 facets each, and their flying speeds of up to 20 miles (32.2 kilometers) per hour! The adults of many species of dragonfly often range far from water. I have seen a dragonfly that was .5 mile (.8 kilometer) from the nearest pond as it hovered outside a ninth-story window, then flew away over the building! Males of many species, like the white-tailed skimmer, guard their mating grounds and egg-laying females.

A common prey for the odonates is the water mite, an eight-legged arachnid related to spiders and scorpions. Two tiny eyes help them to locate their prey—small worms, insects, and crustaceans—as they swim or crawl about during the day. Some water mites are parasitic.

FIGURE 3–49: Water mite, *An arachnid.*
Size: .1–.2 inch (2.5–5.1 millimeters).

Winkle
(Snail)

Pill clam

FIGURE 3–50: Mollusks are abundant on and
in the pond bottom. Winkle (Snail).
Size: 1.5 inches (3.8 centimeters). Pill clam.
Size: .3 inch (7.6 millimeters).

Several kinds of mollusks are found in the pond. Snails (*gastropods*) abound here, scraping algae off rocks and plant stems, and sometimes eating dead animals with their rasplike *radulae,* containing thousands of teeth. They have an internal gill and two eyes, each at the base of a sensory stalk. Gelatinous masses of snail eggs are common on rocks and plants. Another mollusk, the pill clam (*Pelecypoda*), is the size of a hole in a three-ring paper punch. While resting on the bottom the pill clam siphons algae and plant debris from the water while absorbing oxygen through its gills.

A few of the larger animals, too, frequent the pond bottom. Catfish, common suckers, and carp scavenge the mud for food. And the winter cold drives turtles, frogs, and others into the mud to hibernate.

ɜ•THE POND THROUGH THE SEASONS•ɜ

Late winter is a good time to begin following the pond's yearly cycle. Life in the quiet waters begins to awaken. The sun is arching higher as March comes to a close; it casts warming rays onto the slushy ice that is piling up on the leeward shore of the pond. Water from the melting ice sinks until it reaches 39.2°F (4°C), then it rises once again, creating thermal currents that, along with the wind, stir up the pond. This *spring overturn* brings phosphorus and other nutrients up from the bottom and mixes them into the open water. (See Fig. 3–51, p. 106) *Melosira* is one of the diatoms that thrive in the nutrient-rich ponds of spring. Dissolved oxygen levels can climb to 14 ppm, near saturation, during the daylight hours when these cool waters are whipped by the wind and diatoms are active in the sunlit water.

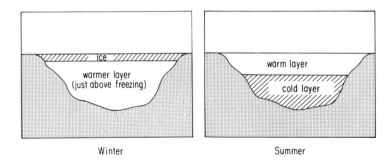

Winter   Summer

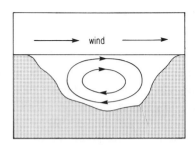

Spring and fall overturns

FIGURE 3–51: The pond through the seasons. Thermal stratification occurs during the winter and summer in temperate
regions. Spring and fall are times of thermal mixing.

FIGURE 3–52: A sign of spring's arrival: the male red-winged blackbird.

Soon the days are lengthening. Frost no longer greets the red-winged blackbirds, tree swallows, mallards, and other early spring migrants to the pond. Male redwings can be seen in a feeding frenzy as they burst into swarming clouds of newly hatched midges and eat them by the thousands. The female redwings will return north later to find a mate who will already have staked out his territory. Insect eggs are hatching in the warming muds along the shore, and water lily roots are sprouting new shoots. Wood frogs can be heard in great numbers, sounding convincingly like a small raft of ducks. Spring peepers turn the evening hours into a chorus of sleighbell-like calling that cannot be ignored. The mating rhythms of springtime bring the pond to life.

Spring ages, and the shoreline waters are crowded with insects: Some are swimming, some are flying, and most are mating. Two metallic-blue damselflies skirt the pond's surface together as the female dips her abdomen into the water to deposit fertile eggs.

June arrives, and life in the pond begins to hit its stride. Mayflies are hatching by the hundreds at the surface, only to be snapped up by swallows darting by. The surface of the pond is now warmer than the deeper waters for much of the day. In an aggressive strike, a newt wrestles a frog-egg meal from its gelatinous coating. All of this life begins to deplete the nutrients in the still waters, and oxygen levels dip low during the short nights. A tiny light can be seen in the darkness coming from the mud along the shore. It is the glowworm, larva of the firefly (which is really a beetle). The larva and adult both combine an enzyme, *luciferin*, with oxygen to create light that is virtually without heat.

By July the fruits of the spring are evident along the shoreline. The water is cloudy with algae, microscopic animals, and other life. Schools of inch-long bullhead fry are closely guarded by their parents in the shallows. Tiny American toads only one-third inch long swarm over the shore, looking like brown pebbles leaping into the air. Mallard ducklings obediently follow their mother in single file as they feed on duckweed and submerged plants in the shallows. A snapping turtle lurks in the weeds, waiting for a tasty, feathered meal.

A contemplative mood overtakes the pond in August and late summer. Water lilies open each morning and close by midafternoon. The water turns green with a dense growth of algae. An occasional bullfrog or green frog bellows out a single call, momentarily shattering the sunlit stillness. Velvety cattail flowers are maturing. Splat! A bass falls back into the water after leaping for a hapless insect. The pond plants and animals seem to be waiting, finishing up their vital tasks, as the first whisper of fall rolls over the pond with each cool morning mist.

One day in September the pond awakens and the swallows have gone south during the night. Sensitive ferns stand out along the shore as crinkly brown leaves bear testimony to the first frost of the season. Soon water lilies are breaking up and floating to the shoreline on crisp fall breezes, leaving only the roots to carry on through the winter. Algae are leaving their resting stages in the muck, as are rotifers and water fleas. Smaller populations of these plankton, along with cyclops, will be active all winter long. Frogs and turtles are spending more of their time in the mud during the increasingly cold nights.

Eventually the pond water approaches freezing. As it cools to 39.2°F (4°C), it sinks to the bottom, then rises to the surface again upon further cooling. This thermal mixing, which is enhanced by the late fall winds, causes a thorough *fall overturn*. Nutrient levels are high in the winter pond, but there are few plants or animals active to use them.

FIGURE 3–53: The pond sleeps. (*Photo by Michael J. Caduto.*)

On many mornings now there is an encrusting layer of ice that rings the shoreline. Gradually the open water shrinks and yields to the sheet of clear, black ice of early winter. The pond is not entirely asleep. An occasional turtle can be seen through the ice, and many organisms carry on for the winter with reduced activity. If the black ice becomes thick enough to walk on, it forms a perfect observation post, acting as a giant diving mask to clearly reveal the pond's secrets. Many algae are still growing, some fish dart by, and a muskrat veers away from your shadow as it hurries past. Cold and frustrated, a belted kingfisher sits on an overhanging branch, able to see its meal but not to catch it! For a time it will eat berries, insects, and even mice to get by. Shortly, it too will fly south to seek open waters. The pond will wait quietly, under a blanket of snow and ice, for the return of the kingfisher's raucous rattle in the springtime.

❧ EXPLORATIONS AND ACTIVITIES ❧

A pond, like any other natural area, is home for plants and animals. Please try to keep disruption to a minimum. Look briefly at the animals and, if you are not going to bring them home for further study in an aquarium, release them unharmed. Leave the environment as close to its original state as possible.

AQUARIUM: A MINI-POND

Establishing an aquarium in which to keep pond life is a fascinating way to learn about animals and their behavior. Many pond animals will keep well even without aeration if the water is clean and has plants growing in it, but fish will require an aeration device.

Begin with a base of pond sand laid over gravel, plant some sub-mergents—common elodea does well here—and put some snails in to graze on algae that may grow on the bottom and sides of the tank. Later you can add insects, frog eggs, small turtles, frogs—whatever you can find. But first learn about your intended catch and how to care for it. Do not overstock the aquarium, and be sure there is a dry place for air-breathing creatures to climb out of the water. Be careful which animals are placed together: Large predators will eat their smaller tank mates! Turtles and fish will thrive on commercial fish food, but you will need to catch the right food for other animals. Rotate your animals by releasing them after a week or two and restock them with replacements to assure a healthy mini-pond and the longevity of the animals.

MAP YOUR POND

With a boat or canoe, you can cross the pond on imaginary lines, as established on shore, and check the depth at intervals. This will give you a profile of the pond bottom. Locate significant plant communities and zones on your map, and mark any sunken logs and large rocks or other objects. Note the source for the pond and its outflow. (See Fig. 3–54, p. 110)

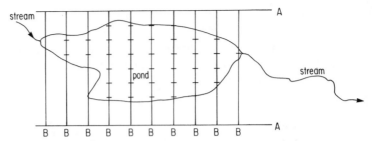

FIGURE 3–54: A method for mapping a pond. A = Lines established and marked at intervals running along the long axis of the pond. B = Imaginary transect lines to follow while crossing the pond during mapping. Cross markings represent points along these lines where measurements and observations can be made: for example, depth, plant communities present, bottom type, and animals seen.

POND JOURNAL

A systematic and accurate journal is one of the best learning tools for studying natural history in the field. Visit the pond on a regular basis—for instance, one time each week—and observe the animal activity, changes in plant growth, and temperature of the water, air, and weather. Record your findings in a field journal. Later, you can file them on 3-by-five-inch note cards, using one card for each plant or animal. You will begin to see patterns in the growth and behavior of organisms in your pond. Photographs and drawings are excellent for recording pond life through the seasons.

POND SAMPLING AND COLLECTING

The basics of pond study are simply to keep alert for the unusual event while patiently sieving the muck and searching the shore for plants and animals. It is not necessary to kill and collect specimens in order to learn about them; they are better teachers if studied in their homes or in an aquarium. Most of the following equipment can be gathered or made from common household items.

- *Old sneakers.* A rule of thumb is that your feet will get wet!
- *Hand lens.* Inexpensive, plastic hand lenses can usually be bought from a biological supply house or nature center. These are a must to view the small pond creatures. Hold the lens up to your eye and adjust the distance of the object until it is in focus. This will give you a more steady view with a wider

field of vision than the Sherlock Holmes approach, where the lens is held at a distance.

- *Terrestrial insect net.* Many adult forms of aquatic insects can be caught by swishing this net through the shoreline vegetation.

- *Mud strainer and Aquatic net.* A large, metal food strainer, with an opening of 6 inches (15.2 centimeters) or greater, makes an excellent collector and sieve for searching the sediments. If fitted with a long handle, it can be used for reaching deeper water from the shore.

- *Plankton net.* The net-shaped tip of a nylon stocking can be used to create a rough plankton-collecting device. This can be dragged behind a boat to collect in the open water. (See the activity section at the end of Chapter 4 for details on making a plankton net.)

- *Pond scope.* Three empty frozen juice cans and a piece of clear acetate are all you need to make a pond scope. Take the bottoms off the empty juice cans and fasten them into a long tube using strong, waterproof tape. Then glue the acetate onto one end. Submerge the end with the acetate attached and sight down the inside of the tube, being careful not to get any water inside. You can now see clearly as you probe the pond bottom. In a pinch, a diver's face mask will do. Just hold it in the water, glass side down, being careful not to spill any over the sides.

- *Thermometers.* Two thermometers are best: one for the air temperature and one for the water. Or you can get an indoor/outdoor model and take both at once. The kind of thermometer with a remote sensing tip is convenient for measuring temperatures at different depths.

- *Yardstick or measuring tape.* A rule is always handy for measuring depths and the size of large organisms. Nylon cord can be fitted with a weight and marked at measured intervals with waterproof ink to make an efficient sounding device.

- *Collecting pans.* The best kinds are white enamel or plastic pans, which form a high-contrast viewing background. Keep the pans nearby in the shade along the shore and store your organisms in several inches of water.

- *Chemical testing kits.* These are expensive but valuable to the avid student of aquatic ecology. Kits can be ordered to measure the level of virtually any freshwater gas or compound such as dissolved oxygen, carbon dioxide, pH, phosphate, and nitrate. Two of the major companies carrying such kits are the Hach Company and LaMotte.

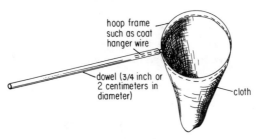

hoop frame such as coat hanger wire

dowel (3/4 inch or 2 centimeters in diameter)

cloth

FIGURE 3–55: Homemade terrestrial insect net.

&ADDITIONAL READING&

Amos, William H. *The Life of the Pond*. New York: McGraw-Hill Book Co., 1967.

Andrews, William A. *Freshwater Ecology*. Englewood Cliffs, N. J.: Prentice-Hall, Inc., 1972.

Burt, William Henry. *A Field Guide to the Mammals*. Boston: Houghton Mifflin Co., 1976.

Cobb, Boughton. *A Field Guide to the Ferns*. Boston: Houghton Mifflin Co., 1963.

Coker, Robert E. *Streams, Lakes, Ponds*. New York: Harper and Row, Publishers, 1968.

Conant, R. *A Field Guide to the Reptiles and Amphibians*. Boston: Houghton Mifflin Co., 1958.

Edmondson, W. T., ed. *Freshwater Biology*. New York: John Wiley and Sons, Inc., 1959.

Farb, Peter. *The Insects*. New York: Time-Life Books, 1962.

Frost, S. W. *Insect Life and Insect Natural History*. New York: Dover Publications, Inc., 1959.

Gravé, Eric V. *Discover the Invisible: A Naturalist's Guide to Using the Microscope*. Englewood Cliffs, N. J.: Prentice-Hall, Inc., 1984.

Hotchkiss, Neil. *Common Marsh, Underwater and Floating-leaved Plants of the United States and Canada*. New York: Dover Publications, Inc., 1972.

Hutchins, Ross E. *Insects*. Englewood Cliffs, N. J.: Prentice-Hall, Inc., 1966.

Jahn, T. L., and F. F. Jahn. *How to Know the Protozoa*. Dubuque, Iowa: William C. Brown Co., 1949.

Keeton, William T. *Biological Science*. New York: W. W. Norton and Company, Inc., 1972.

Klots, Elsie B. *The New Fieldbook of Freshwater Life*. New York: G. P. Putnam's Sons, 1966.

Margulis, L., and K. V. Schwartz. *Five Kingdoms: An Illustrated Guide to the Phyla of Life on Earth*. San Francisco, California: W. H. Freeman Co., 1981.

Newcomb, Lawrence. *Newcomb's Wildflower Guide*. Boston: Little, Brown and Co., 1977.

Niering, William A. *The Life of the Marsh*. New York: McGraw Hill Book Co., 1966.

Ommanney, F. D., et al. *The Fishes*. New York: Time Inc., 1964.

Orr, Robert T. *Vertebrate Biology*. Philadelphia: W. B. Saunders Co., 1976.

Pennak, Robert W. *Freshwater Invertebrates of the United States*. New York: John Wiley and Sons, Inc., 1978.

Prescott, G. W. *How to Know the Aquatic Plants*. Dubuque, Iowa: William C. Brown Co., 1969.

_____. *How to Know the Freshwater Algae*. Dubuque, Iowa: William C. Brown Co., 1970.

Reid, George K., and Herbert S. Zim. *Pond Life.* New York: Golden Press, 1967.
Zim, H. S., and H. H. Shoemaker. *Fishes.* New York: Golden Press, 1955.

❧*NOTES*❧

1. Ralph W. Tiner, Jr., *Wetlands of the United States: Current Status and Recent Trends* (Washington, D.C.: United States Fish and Wildlife Service, 1984), p. 31.
2. L. Margulis and K. V. Schwartz, *Five Kingdoms: An Illustrated Guide to the Phyla of Life on Earth* (San Francisco, CA: W. H. Freeman Co., 1981).

CHAPTER FOUR

LOOKING AT LAKES

Eleven canoes emerged from the peaceful waters and cattail marshes at the mouth of Dead Creek. The calls of black ducks and kingfishers still echoed in our ears. Tall rushes cast an unbroken reflection across the flowing, flat waters, sliced by the prows of canoes under power of tired arms. Buffeted by the sudden head winds accelerating across the broad fetch of Lake Champlain, we tried to outdo the fury, gust for paddle stroke. Towering thunderheads rode toward us on the winds, urged on by white-capped waves breaking over our gunwales. Canoes began filling from rain and lake.

FIGURE 4–1: (*Photo by Cecil B. Hoisington.*)

"Over there," someone cried out. "They've tipped over!"

Two people up ahead were clinging to a swamped canoe. We in the lead canoes paddled hard upwind, adrenaline rushing, a dry feeling in our throats. Was this the often-feared moment of disaster that can turn a pleasant excursion into a drama read in tomorrow's news?

"Hey, we can stand up!" yelled one of the waterborne canoeists. "We'll be all right, go on in to shore."

The rain slowed, then stopped. We reached, then skirted the rocky shoreline, passed an island, and soon stepped ashore in the calm, clear waters of an inlet. Anxious expressions melted into smiles of relief, and finally the shrieks and thrills of swimming people filled the air.

❧FROM POND TO LAKE❧

There are times when there is a distinct line between a calm, pondlike environment and a lake, and this was clearly one of those times. Recall the character of a pond as described at the beginning of Chapter 3. While sometimes it can be as shallow as a pond, the leeward shore of a lake is molded by the action of wind and waves. Fewer plants grow on the sandy, rocky, wave-beaten shores than in the sheltered coves. Attached algae commonly coat the rocks here. Out in the deep water near the lake bottom there is not enough sunlight penetrating to support plant growth. The dissolved oxygen levels and temperatures are relatively stable in this dark world com-

FIGURE 4–2: The leeward lakeshore—windblown and wave-beaten. (*Photo by Michael J. Caduto.*)

pared to the fluctuation in the shallow open waters above. Even at the surface of a lake, the daily changes in DO levels and temperature are less marked than in a pond. A warmer layer, which persists throughout the spring, summer, and fall, forms over the cold water below. And because of the larger size and depth of lakes, natural lake succession, while it is occurring, must be thought of in terms of thousands, and—in the case of very deep and oligotrophic lakes like Lake Baikal in Russia—even millions of years.

❧LAKE ORIGINS❧

Lakes are created by some of the same forces that form ponds, but on a much larger scale. There are also some more powerful agents at work in the formation of lakes. The shape, size, and surrounding geology and environmental conditions determine the individual character of a lake.

GLACIERS

In glaciated regions, magnificant depressions were formed, which then filled with glacial melt water. The Finger Lakes in New York; Lake Champlain, lying between New York, Vermont, and Quebec; and the Great Lakes of the Midwest are examples. Many of the thousands of smaller lakes in the northern areas are also a legacy of the glacier.

TECTONIC FORCES

The movement of the earth's crustal plates as they rise, sink, and drift—together in some places and apart in others—has formed enormous lakes.

FIGURE 4–3: Crater Lake (Oregon) formed in the caldera of Mount Mazama which erupted nearly 7,000 years ago. (Photo by Michael J. Caduto.)

Lake Okeechobee in Florida was isolated from the sea by the rising of the crustal plate upon which it rests. Tectonic forces formed Lake Tanganyika in east-central Africa, with its 12,700 square miles (5,140 hectares) of open water; an area 1.5 times the size of Massachusetts. Some of the largest, deepest, and clearest lakes in the world are of tectonic origin.

VOLCANOS

The craters or *calderas* left after a volcano spews forth its load of lava over the landscape can fill with water to form very clear, oligotrophic lakes. Crater Lake in southwestern Oregon, for example, is 6 miles (9.7 kilo-meters) long, 5 miles (8 kilometers) wide, and around 2000 feet (610 meters) deep; it is considered one of the clearest lakes in the world. Two reasons for the clear waters of volcanic lakes are (1) the hard volcanic rocks over which the lakes form, which contribute few nutrients to the surface runoff that enters the lakes, resulting in soft waters that are low in fertility; and (2) the relative youthfulness of these lakes and their watersheds, many of which were formed in recent geologic times. There is little or no soil development or plant growth to contribute nutrients to the surface runoff that enters the lakes, nor can these areas support agriculture or other human activities that might add nutrients to the water.

RIVERS

Besides scroll and oxbow lakes and ponds, larger *delta lakes* can form at the mouth of a river when sediment loads form natural dams and levees. *Plunge pool lakes* are carved at the base of large waterfalls.

SOLUTION LAKES

Deep Lake in Florida is an example of a lake resting in a depression created by the solution of soft rock and the resulting subsidence of the overburden.

PEOPLE

Although many small ponds and lakes are created by people, some of our larger lakes rest behind dams. The Quabbin Reservoir in central Mas-sachusetts, a major water supply of the city of Boston, is almost 20 miles (32.2 kilometers) long. The damming of many rivers for recreation, hydro-lectric power generation, and water supplies—like the Connecticut River, which flows from Quebec to Long Island Sound—has turned them into a series of long lakes. Dams along the Colorado and Columbia rivers in the western United States have created very large lakes. Lake Mead, which rests behind the Hoover Dam along the Colorado River, stretches over 100 miles (161 kilometers) through the desert and is 8 miles (13 kilometers) across at its widest points. With a surface area of 250 square miles (101 hectares), Lake Mead is almost one fifth the size of Rhode Island. Hoover Dam itself is 1244 feet (379 meters) long and 726 feet (271 meters) high!

THE LAKE ENVIRONMENT

When spring arrives in the lake, as in a pond, the upper layer is churned by the wind and thermal convection, mixing gases (from the air and photosynthesis) and nutrients (from the bottom and from incoming runoff) into the water. In many lakes this mixing occurs down to the bottom. The spring overturn spreads gradually from the shore, where the shallower water warms more quickly, and out into the open water. By early summer, several distinct thermal layers have formed. (Recall from Chapter 1 how cold water—down to 39.2°F or 4°C—is more dense and sinks to the bottom.) In the upper layer or *epilimnion*, heating, cooling, and wind create strong currents to a depth of up to 66 feet (22 meters). Below this layer is the *metalimnion* or *thermocline*, an area often only a few meters deep, which forms a transition zone between the active epilimnion and the colder, still waters or *hypolimnion* below, which does not experience thermal mixing during the summer. Most temperate lakes are *dimictic*, experiencing an overturn in the spring and fall, while in more southerly, *monomictic* lakes, the epilimnion mixes constantly during the winter. The formation of these vertical temperature zones in lakes is known as *thermal stratification*.

The inertia of the earth's rotation causes the *Coriolis force*, which creates a slow, steady current in very large lakes: counterclockwise in the Northern Hemisphere and clockwise south of the Equator. This movement combines with thermal mixing and wind-generated currents to form the pattern of water movement in a lake. In addition, the wind and waves move large volumes of water toward the leeward shore. This water pushes down into the hypolimnion, which establishes a deeper current flowing back toward the windward shore. A rocking motion results in the lake, somewhat like the back-and-forth movement of water that occurs in a bathtub when you slide from end to end. This motion adds another rhythm in the dance of the lake waters.

Moving from shore, the major habitats of the lake are the *littoral* zone, which has as its furthest extent the limit of rooted plant growth, and the *limnetic* or *pelagic* zone—the open water. The photic zone is that water lying

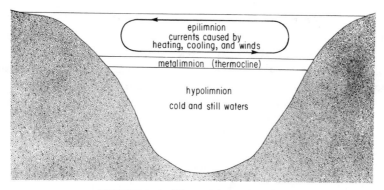

FIGURE 4–4: Thermal layers in a lake.

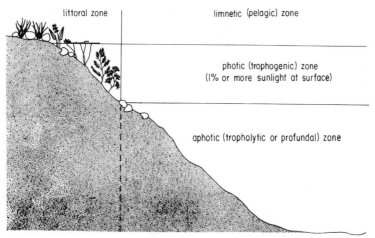

FIGURE 4–5: Zones of a lake as determined by sunlight penetration.

above the point where the light level is 1 percent of the available sunlight at the surface. Because there is sufficient light here to support the growth of planktonic algae, this is called the *trophogenic* or nutrient-producing zone. Below this lies the *aphotic* ("no light"), *tropholytic*, or *profundal* zone, where no plants can grow. The epilimnion forms predominantly in the photic zone.

Thermal stratification determines the plant and animal communities by the different levels of nutrients and dissolved gases in each zone. The epilimnion is a zone of growth where plants photosynthesize and produce oxygen during the day. Here, the daily cycle of oxygen levels is similar to that in a pond, but is less exaggerated. The thermocline sometimes has the greatest DO levels in a lake when it contains high concentrations of photosynthesizing algae. In some very clear, unproductive lakes, sufficient sunlight may penetrate to support phytoplankton in the hypolimnion, and oxygen levels may even be higher there than at the surface. During the summer growing season in temperate climates, fertile lakes experience high production rates in the epilimnion. As a result, because the plants and animals that are produced eventually die and sink into the hypolimnion, where they decompose, oxygen levels decrease gradually with increasing depth in the hypolimnion of productive waters. The thermocline, then, is a major force determining conditions for plant and animal growth and occurrence. The *compensation depth* is the vertical zone where oxygen production equals respiration levels.

Suppose you were an alga living in the epilimnion. How would thermal stratification affect your food supply throughout the summer? Early in the growing season, right after the spring overturn, nitrogen, phosphorus, and other nutrients would be abundant. Much of the nitrogen present had built up in the water during the winter from precipitation and runoff, which carried nitrogen-containing sediments into the lake, as well as from the nitrogen already contained in lake sediments. Phosphorus was brought up from the bottom during the spring overturn or entered the lake in surface runoff. Gradually, as summer progresses and you and the other algae grow, these nutrients are depleted. Phosphorus becomes in very short supply, but you and some other algae have stored phosphorus for use during this shortage. Nitrogen fixation helps some algae to supplement their supply of that nutrient.

As the waste products and remains of dead organisms slowly settle into the hypolimnion, phosphorus and nitrogen compounds become concentrated in the deep water. The organic phosphorus is decomposed by anaerobic bacteria, which can change it to a form once again usable by plants at a rate one thousand times faster than the microbes living in aerobic conditions! When the fall overturn finally comes, more nutrients are brought up from the bottom, and there may be a brief algal bloom before the winter arrives. Nutrient levels will be high during the winter.

Besides the effects of seasonal changes, it is important to recall the many other natural influences on nutrient levels and water chemistry in lake waters. These include the solution of bedrock and soils by ground water and runoff, local climate, plant cover, topography of the watershed, and inputs from precipitation. Detailed discussions of these influences are found in Chapters 1 and 2.

HUMAN INFLUENCE ON LIFE IN LAKES

The effects of human activities on water quality and on the abundance and health of lake plants and animals is a matter of increasing concern. Chapter 2 discusses in depth the ecological changes wrought in lakes by the introduction of excess nutrients, toxic elements such as acid rain, and the physical alteration of still water environments and their surroundings. This discussion is continued in a closer look at lake plants and animals.

❧ LAKE PLANTS ❧

Plants can be found in the littoral zone and in the open water where sunlight is adequate to support photosynthesis. In sheltered shoreline areas, the shallow water may harbor some species described in Chapter 3 as part of the pond environment, as well as many plants that are common in lakes. The reddish flowers and greenish-purple stems and leaves of marsh Saint Johnswort are often seen along the shore. Its hardy seeds sprout up in the rotting tops of partially submerged stumps and logs in sheltered coves. Water parsnip, a white-flowered member of the parsley family, frequents the protected lakeshore. A plant called quillwort, a close relative of the clubmosses, is a casual sighting in the shallow waters. (See Fig. 4–7, p. 122)

Chairmaker's rush or threesquare, which is really a sedge, may be seen

FIGURE 4–6: Marsh Saint Johnswort (*Hypericum virginicum*). *Size:* 1–2 feet high (.3–.6 meter). (*Photo by Michael J. Caduto.*)

FIGURE 4–7: (*left*) Water parsnip (*Sium suave*). *Size:* 2–6 feet (.6–1.8 meters); (*right*) Quillwort (*Isoetes spp.*). *Size:* 3–20 inches (7.6–50.8 centimeters).

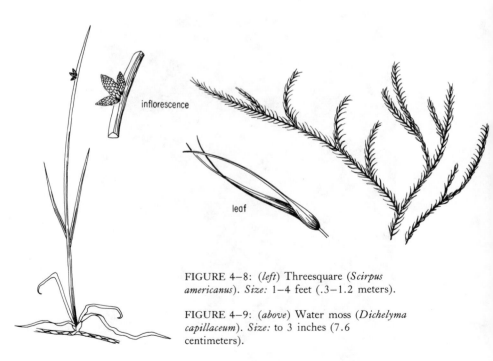

inflorescence

leaf

FIGURE 4–8: (*left*) Threesquare (*Scirpus americanus*). *Size:* 1–4 feet (.3–1.2 meters).

FIGURE 4–9: (*above*) Water moss (*Dichelyma capillaceum*). *Size:* to 3 inches (7.6 centimeters).

where the waves are more active. These three-sided stems are so tough that they were once used for caning to weave the seats of chairs. Growing in a more sheltered shallow zone, but with a similar growth habitat of single, stalwart stems swaying in the waves, is the soft rush. Water moss may be draped over a submerged rock or log nearby. A submergent plant, naiad, can grow in long, billowing tufts along the lakeshore.

Planktonic algae are the major photosynthetic producers in the limnetic zone. (Fig. 4—10, p. 124) There are also some bacteria found here that are photosynthetic, as well as other bacteria that are *chemosynthetic*, being capable of growing and creating organic matter from inorganic substances without the aid of sunlight. Algae are more abundant during the warm seasons of spring and summer when nutrient levels, available sunlight, and water temperatures are higher than during the rest of the year. The numbers of algae and their distribution within a lake are also affected by their movements, competition with other algae for available nutrients, attack by disease-causing organisms, and the consumption of algae by herbivores such as filter-feeding zooplankton.

The same major groups of algae are found in lakes as in ponds, and the reader will find additional information and illustrations under the descriptions of pond algae.

Diatoms lend their beautiful designs to the community of lake phytoplankton. *Fragilaria* and *Tabellaria* are common, *Tabellaria* with its zigzag chains that look like miniature renditions of Jacob's ladder. Hardwater lakes often harbor healthy populations of another diatom, *Asterionella*.

The desmids, such as *Micrasterias*, are common green algae of oligotrophic lakes. *Micrasterias* has a platelike shape with protruding spines; it thrives in soft, acid waters, such as those of bogs dominated by *Sphagnum* moss. *Chlorella* and *Pediastrum* are often collected in lake waters. A symbiotic relationship exists between *Chlorella* and certain species of protozoans, sponges and hydras, with *Chlorella* living inside these tiny organisms. The genus *Pediastrum* is a group of colonial algae shaped like segmented plates, which form colonies ranging from a few to over sixty cells. They are sometimes numerous in very cold waters. See Figure 4—10 on page 124.

Warm, eutrophic lakes can support heavy blooms of blue-green algae. These blooms increase turbidity, which decreases the sunlight reaching the deeper water to support oxygen-producing algae. *Anabaena* causes greenish-gray cloudiness in the water during the midsummer. *Microcystis* forms dense colonies that entrap gas pockets which buoy the masses to the surface as slimy, floating scums. The decay of these colonies can cause a dramatic drop in oxygen levels, sometimes suffocating fish and other aquatic life. Those animals that need especially high levels of oxygen to live, such as lake trout

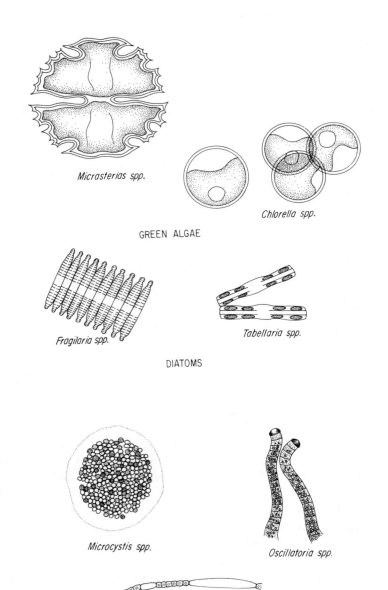

Micrasterias spp.

Chlorella spp.

GREEN ALGAE

Fragilaria spp.

Tabellaria spp.

DIATOMS

Microcystis spp.

Oscillatoria spp.

Aphanizomenon spp.

BLUE GREEN ALGAE

FIGURE 4–10: Some common lake algae. *Size:* Microscopic detail shown.

and landlocked salmon, cannot survive in eutrophic waters. Some species of both *Anabaena* and *Microcystis* form toxins potent enough to kill animals and birds. Your first thought on seeing a floating mass of *Aphanizomenon* might be that it is a mat of weeds or grass. Water containing blooms of *Oscillatoria* takes on a reddish hue. *Oscillatoria* gets its name from the way it moves, with an oscillating motion. These heavy blooms of blue-green algae sometimes create noxious odors. Aesthetically, eutrophic waters are not as pleasant for swimming and other recreational activities as are the clear waters of oligotrophic lakes.

One common organism of the open water, the *dinoflagellate* ("whistling whip") *Ceratium hirundinella*, frequently forms great blooms that turn fertile waters gray to brown. This species has characteristics of both plants and animals, and is sometimes classified as a member of the Protoctista. *Ceratium* has a red eyespot. Two flagella aid in locomotion.

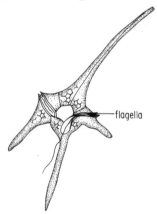

FIGURE 4–11: A dinoflagellate
(*Ceratium hirundinella*) *Size:*
Microscopic detail shown.

≈ANIMALS OF THE LAKESHORE AND BOTTOM≈

One of the most pleasant circumstances for exploring freshwater life is to be wading knee-deep in the clear water along a lakeshore on a hot, sunny day. Predaceous diving beetles will release their holds on submerged plants as they scurry for cover. You may part a path through frantic whirligig beetles, only to see it close again once you have passed. If you are lucky, you may see a colony of bryozoans attached to a submerged log, looking like a miniature reef of some unfamiliar tropical coral.

On passing some rocks you stop to look more closely. You turn away puzzled, then look again. Amorphous and inconspicuous splotches of green and brown cover the rocks, looking almost like large patches of lichen on land. The patches feel somewhat smooth and slippery, and pieces are easily dislodged. This is the freshwater sponge, *Spongilla lacustris*. *Heteromeyenia tubisperma* is another species you may encounter. The sponges, *Porifera*, may be represented by three or four species in a large lake. Some are smooth; others are bumpy and thick. Most are found in water 6.6 feet (2 meters) deep or less. Sponges consist of a series of channels with openings, or *ostia*, through which water, oxygen, and food are drawn by beating flagella. Bacteria, algae, and microscopic animals stick to collar cells inside the sponge and are then digested within the tissues. A skeleton of siliceous spicules lends support to these tissues. The wastes leave via oscula. Reproduction is poorly understood, but usually occurs in July and August. Later in the season a resistant reproductive body, called a *gemmule*, is formed that can withstand cold and drought. The larvae of spongillaflies, *neuropterans*, live in the sponges and eat sponge tissue.

Turning over other submerged objects, you may find a tiny (.4 inch or

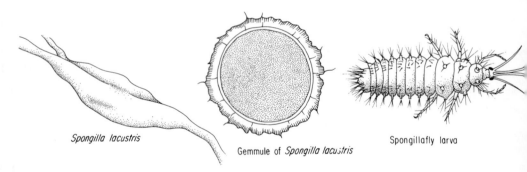

Spongilla lacustris

Gemmule of *Spongilla lacustris*

Spongillafly larva

FIGURE 4–12: Freshwater sponge (*Spongilla lacustris*). *Size:* Variable, colonies are around 7 inches (17.8 centimeters) across. Gemmule of *Spongilla lacustris*. *Size:* Microscopic detail shown. Spongillafly larva. *Size:* .3 inch (7.6 millimeters).

1 centimeter) *Hydra*. These *hydrozoans*, close relatives to the jellyfish, are most common in late spring and early summer in the clear waters of lakes and rivers with sandy or rocky bottoms. They prefer hard, slightly alkaline waters. When a small crustacean or insect happens by, one or more of the *Hydra's* six or more hollow tentacles springs out and stings the prey, paralyzes it, and entraps it. Digestion takes place in its hollow, stalklike body. Some *Hydras* reproduce by budding, while others mate. Also part of the neuston community, a *Hydra* can rise up on a bubble and hang from the surface film. The common green *Hydra* has commensal algae livng in its body cells.

Water fleas and copepods, tiny crustaceans that are among the *Hydra's* staples, are abundant along the shore. Some species of copepod occur out into the limnetic waters and, along with water fleas, are a major component of the pelagic zooplankton.

A long, fleeting shadow tells you that a pickerel has just been spooked and has fled. With sharp teeth and fast charges, this fish feeds on frogs, other fish, insects, and even an occasional duckling amid the shoreline vegetation. It is a small relative of the larger pike of the open water.

The littoral bottom is alive with many animals also found in the pond, depending on the specific conditions of plant growth, wave action, bottom type, pH, and nutrient supply. As depth increases and the bottom stretches into the hypolimnion, light levels and temperatures drop. Dissolved oxygen

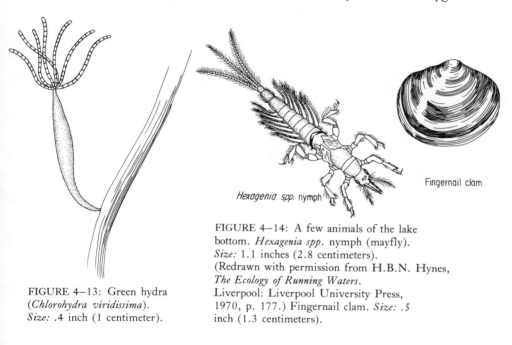

Fingernail clam

FIGURE 4–14: A few animals of the lake bottom. *Hexagenia spp.* nymph (mayfly). *Size:* 1.1 inches (2.8 centimeters). (Redrawn with permission from H.B.N. Hynes, *The Ecology of Running Waters*. Liverpool: Liverpool University Press, 1970, p. 177.) Fingernail clam. *Size:* .5 inch (1.3 centimeters).

FIGURE 4–13: Green hydra (*Chlorohydra viridissima*). *Size:* .4 inch (1 centimeter).

levels may decrease with increasing depth during the summer in eutrophic waters. The benthic community of the deep water is limited in the number of species because of these environmental extremes, but many individuals of some species may be found. Bloodworms, the larvae of true midges, are common, as well as the transparent phantom midge larvae. There may be over fifty species of midges living in one lake! Occasionally great numbers of the mayfly *Hexagenia* may occur beyond the littoral zone, along with the fingernail clams of the family *Sphaeridae*. In the deepest waters, anaerobic bacteria, blood worms, and tubifex worms are the chief denizens. Most bottom-dwelling animals of the deep waters feed on detritus and in turn become food for fish and other animals.

❧ANIMALS OF THE OPEN WATER❧

The most common zooplankton of open lake waters belong to similar groups to those described and illustrated in the section of Chapter 3, Animals of the Open Water. A planktonic sample of most lakes will reveal several species of rotifer and a few species each of two major groups of crustaceans—water fleas and copepods. Protozoans are also common.

The chief foods of zooplankton are bacteria, algae, detritus, and other, smaller zooplankton. Populations of zooplankton rise and fall in part as a result of the available food supply. The water fleas, *Daphnia*, have resting stages that can develop quickly when conditions become favorable, such as during algal blooms. Consequently, zooplankton are more numerous in fertile waters where food is plentiful. Water fleas will produce resting stages when food supplies become scarce.

Specialized adaptations also help these tiny animals to survive the changing conditions of predation pressures, temperatures, nutrient supplies, competition for available food, and other factors such as fluctuating dissolved oxygen levels. Fish eat many of the larger crustaceans, such as certain species of *Daphnia*. Smaller zooplankton are eaten by immature fish and other predators. But zooplankton are also well adapted to avoid being caught by predators that hunt by sight. Some common strategies include small size, nearly transparent bodies, and movement into deeper, darker waters during the day to keep from being seen.

As in ponds, water fleas, rotifers, and copepods are common during the summer months, while copepods are active in large numbers during the winter.

Zooplankton fall prey to the plankton-eating young stages of almost all lake fish of the open waters. Among those fish that eat plankton throughout their lives are smelts, shad, herring, lake whitefish, sunfish, young bass, and numerous minnows.

These plankton eaters, and the bottom-feeding animals that devour small benthic life such as insects and worms, are in turn food for the large, carnivorous fish for which lakes are known. The names of northern pike, muskellunge, lake sturgeon, and lake trout elicit images of large, long-lived, and respected bony fishes. While you may sometimes see one of these lunkers lurking along the lakeshore shallows in the springtime, they spend most of the year in the chill of deep waters.

It is not uncommon for lake trout to live twenty years, and some have been found that were estimated to be forty years old! They can grow to gargantuan proportions and up to 100 pounds (45 kilograms), but are usually not much larger than 24 inches (61 centimeters) and 11 pounds (5 kilograms). Lake trout are prized in sport fishing. Deep-water ledges are one of their favorite summer haunts, although they migrate up to the surface during the spring overturn.

Landlocked salmon, members of the trout family, can grow to be about as large as lake trout, and they share similar habits. These beautiful and legendary fish look like their saltwater cousins the Atlantic salmon (illustrated in Chapter 5). They prefer deep waters, around 33–51 feet (15–23 meters), and temperatures of 50–54°F (10–12°C). Spring is an active season when they surface to feed on smelts and other small fish. Spawning takes place during the fall in the gravels of rivers and streams that feed the lake.

FIGURE 4–15: Lake fish of the open water.

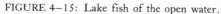

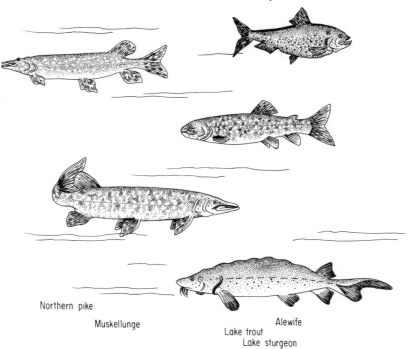

Northern pike

Muskellunge

Alewife

Lake trout

Lake sturgeon

Of all the creatures of the lake, few are so surrounded by mystery as the muskellunge. Muskies are primarily fish of the northern lakes. The largest individuals have been recorded at up to 100 pounds (45 kilograms) and 8 feet (2.4 meters) long. Its lone habit and strongly territorial nature have contributed to the respect with which the muskellunge is regarded. Small mammals, young waterfowl, and smaller fish are all food for the muskellunge.

The northern pike is a close relative. A popular game fish, it is found amid the weeds in shallow lakeshores, ponds, and rivers. The pike's large mouth is filled with sharp teeth, which are wielded with lightning speed when it charges. Northern pike seem to be all mouth at first glance, with a head that is shaped like a duck's bill.

Although the alewife is more commonly known for its *anadromous* habits, living in salt waters and spawning in fresh, they occur as landlocked and introduced schools in many lakes. Similar in appearance to the American shad of the Atlantic coast, the alewife can grow to be 10 inches (25 centimeters) long and up to 1 pound (.45 kilograms).

Another large lake fish, the lake sturgeon, is virtually absent from its ancestral lakes due to overfishing and the damming and pollution of the clean rivers it needs to spawn. Its demise has also been hastened by its slow growth rate. The sturgeon is a primitive-looking fish which has large bony plates covering the head when young. These plates get smaller with age and nearly disappear in older fish. Sturgeon can attain a length of 6.6 feet (2 meters) or more, and weigh well over 100 pounds (45 kilograms).

Sea lampreys, which are cyclostomes or jawless fishes, are now a common member of the Great Lakes due to the construction of canals connecting the sea to these inland lakes, giving the lamprey access that it previously lacked. Lake Ontario was a natural home for the lampreys, with access from the ocean through the Saint Lawrence River. When the Welland Canal was completed in 1892, the lamprey gained access from Lake Ontario into Lake Erie. Before this time, Niagara Falls formed a natural barrier preventing the migration of sea lamprey into the other Great Lakes. The sea lamprey gradually made its way into Lake Huron, and later into Lakes Michigan and Superior. Heavy commercial fishing and the parasitic habits of the lamprey both contributed to the decline of salmonid populations in the lakes, especially lake herring, lake trout, and lake whitefish. A female lamprey can lay fifty thousand eggs in one clutch. Those that survive will grow into adults bearing a sucking mouth with 125 teeth arranged like seeds on the head of a sunflower. These teeth will be used to rasp a hole in the side of its host, from which body fluids will be drawn until the host is near death. The lamprey then moves on to find another host.

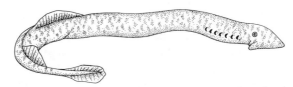

FIGURE 4–16: Sea lamprey (*Petromyzon marinus*), a parasite that has decimated the populations of many fishes of the Great Lakes.
Size: to 3 feet (.9 meter).

The American eel is a little-known creature of the lake. It spends most of its adult life in a freshwater lake or pond, where it feeds along the bottom. When it reaches the age of from five to eight years, the mature eel leaves its native waters to spawn in the Sargasso Sea off the coast of Bermuda. The young elver finds its way to its ancestral freshwater home, where it matures, thus beginning the cycle anew. The opposite of salmon and the alewife, eels are *catadromous*, living in fresh water and spawning in the sea.

While eels leave the lake to spawn once in their lifetime, many of the birds of the open water migrate to and from the lake many times. (See Fig. 4−17 and 4−18, p. 132) Hooded mergansers come to nest in hollow trees near the shore. The young feed in the shallows and gradually move out into open water to eat fish and aquatic insects. The eared grebe builds a floating nest attached to plants in the shallows. Unlike ducks, whose feet are webbed to aid in swimming, grebes have separate, lobed toes.

The haunting, eerie wail of a loon may tell you that a nest is nearby. Loon nests are especially fragile because they are built right at the water's edge. A rise in the water level can flood the nest, and too severe a drop can make the nest inaccessible to this bird which, while a powerful swimmer, is nearly helpless on land. When underwater, the loon can maneuver agilely to catch fish, the mainstay of its diet. Small fish are often eaten underwater, while larger ones are taken to the surface, where they are devoured.

Diving birds share some common adaptations that increase their efficiency while swimming underwater. A streamlined body with legs set well to the rear is an ideal arrangement for propulsion and maneuverability underwater. In addition, some birds, notably the cormorants, use their wings for "flying" and steering as they swim. Diving birds do sacrifice mobility on land for their grace in the water, and most need a large area for taking off and landing, often "running" across the water for a distance while taking flight. When a diver goes under, its heartbeat and metabolic rate slow down and carbon dioxide builds up in its respiratory spaces.

During migration, enormous rafts of birds can be seen on the lake: scaups, grebes, common goldeneyes, redheads, and buffleheads. These are just a few of the common migrants that use lakes for migratory resting areas. Some of the herbivorous dabbling ducks and geese of the shallows will remain on the lake late into the fall. The ingenious blood circulation going to their feet provides *countercurrent heat conservation*. The cold venal blood in their feet is in close contact with the warm arterial blood as it comes from the heart. In this way the blood flowing toward the heart is warmed. This prevents excessive cooling of the body while their feet are dangling in water near freezing temperatures.

Bald eagles may be seen around the lake during the winter, when the last patches of open water have shrunk and finally disappeared beneath the

Common loon   ♂ Hooded merganser   Eared grebe

♂ Common goldeneye   ♂ Bufflehead   ♂ Lesser scaup

♂ Redhead

FIGURES 4–17 and 4–18: Diving birds that frequent the open water of the lake.

FIGURE 4–19: The bald eagle is at the top of the lake's food web.
Size: body, 30–43 inches (76–109 centimeters), wingspread, 7–8 feet
(2.1–2.4 meters). (*Photo by Cecil B. Hoisington.*)

encroaching ice, and waterfowl have fled to the south. Fish are the eagle's main food during the warmer months—a prey sometimes taken from the osprey that have already done the hard work of catching the fish. But even when the fish swim safely beneath the lake's icy coat, some hardy eagles find food enough among the remains of unfortunate animals that venture out onto the ice and perish, never to return to their home on land.

Lakes are a composite of different environments—the productive shore zone, where many birds, fishes, and other animals are raised as young; the open water; and the nutrient-rich bottom sediments. Animals living in these environments frequently move from one to the other as they grow. The hooded merganser that is hatched up in a tree cavity will be raised in the shallows, only to take to the open water as it begins to feed on fish and other pelagic animals. Life in lakes is intimately tied to the shallow surrounding marshes and nearby forests. A thread of life runs through these homes, tying all together as one.

❧EXPLORATIONS AND ACTIVITIES❧

Lakes are large bodies of water than can be a challenge to explore without some basic equipment. It is possible to sample the life and living conditions in the shallows using the suggestions and instruments described at the end of Chapter 3. However, a boat will be needed to take measurements out in the open water.

Turbidity and Environmental Conditions

Turbidity indicates the degree of light penetration through the water; it is a measure of the amount of living and nonliving suspended solids, as well as dissolved elements that impart color to the water. The water will become more turbid or cloudy after heavy rains have caused soil and other particles to wash in from the surrounding land, streams, and atmosphere. Algal blooms and dense zooplankton populations also increase turbidity in eutrophic waters, such as those contaminated with organic pollution. Measurements can be taken at regular intervals, following heavy rains, and in areas where erosion is occurring along the shore. Comparative readings from different lakes are useful to gauge how different natural surroundings and human activities can affect turbidity. The general depth of light penetration often corresponds with the limit of plant growth—the extent of the littoral zone.

The Secchi disc (pronounced "seh-key") is used to measure light penetration. It should be lowered slowly until it can no longer be seen, then the depth should be measured. Pull it up until it is visible and take this depth.

Average the two measurements. This should be done on the shady side of the boat.

To make a simple Secchi disc, use either the cover of a half-gallon paint can or a stiff plastic lid of about the same diameter. The black-and-white zones provide the contrast that enhances the visibility of the Secchi disc.

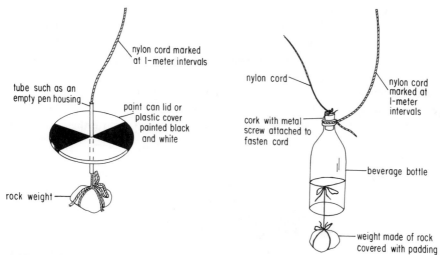

FIGURE 4–20: Parts of the do-it-yourself Secchi disc.

FIGURE 4–21: Deep water sampling bottle.

DEEP-WATER SAMPLING
FOR PLANKTON AND TEMPERATURE

A simple device can be made to take deep-water samples. The sample can be used to test relative clarity of the water (as compared to that at the surface), to take rough measurements of deep-water temperatures, and for observing planktonic organisms found at different depths. Lower the empty bottle to the desired depth, then pull the cork and the bottle will fill with water. Pull the bottle straight up without tipping it, to avoid mixing the sample with other water. Once at the surface, quickly place this sample in the shade and insert a thermometer for an approximate reading of the deep-water temperature.

pH

The pH of a lake is a crucial determinant of the life that can live there. Inexpensive pH papers are available. Concern over the effects of acid rain on water quality has prompted some state, provincial, and private conservation groups to begin extensive voluntary monitoring programs. Contact your local government agency in charge of monitoring water quality to find out if such a program exists. This agency may also be able to provide records of the water-quality testing and biological studies that have been done in your lake.

Plankton Sampling

A simple plankton net can be made using the toe end of a nylon stocking or pair of panty hose. The round support can be made from either a wire frame of the type used for making small Christmas wreaths or the rim cut from the edge of a sturdy plastic lid; or you can fashion your own, using a thin band of sheet metal and screws or rivets. A diameter of around 8–12 inches (20–30 centimeters) is good. Cut the foot of the stocking off about 14 inches (36 centimeters) up from the end. Fold the open part over the rim and sew it securely in place. Attach three leader ropes to the rim and tie these onto the longer rope.

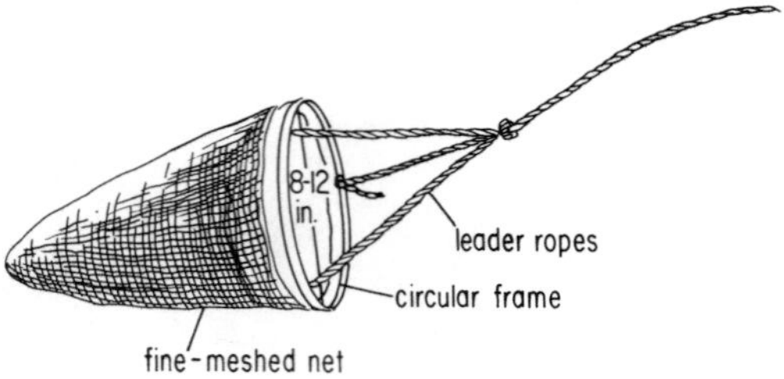

FIGURE 4–22: Plankton net.

This net can be trailed after a boat or dragged by hand. Be careful not to pull it too quickly through the water or the net will split. Remove the plankton by inverting the net into a pan of water and gently swishing the plankton free.

Examining Microscopic Organisms

You will need a microscope to observe the fascinating organisms that are too small to be seen through a hand lens. The details of microscopy, and of what to look for when acquiring a microscope, are too involved to be adequately discussed here. The naturalist is referred to a very comprehensive and useful book that covers all of this material and much more: *Discover the Invisible: A Naturalist's Guide to Using the Microscope*, by Eric V. Gravé, Prentice-Hall, Inc., Englewood Cliffs, New Jersey, 1984.

Assessing the Surrounding Environment

A detailed map can be made of the environs of your lake. Topographic maps reveal all the physical surface features such as roads, buildings, fields, forests, streams, lakes, ponds, and wetlands. The approximate watershed boundaries for the lake can be traced on this map using elevation or contour lines, streams, and river channels. Geology and soils maps will indicate the type of bedrock and soils in the watershed, and whether these conditions may be a source of chemical buffering for the water. Aerial photographs can add to the overall perspective with a detailed picture of the land as it really appears. Permanent stakes can be planted offshore, below the lowest annual water level, to monitor lake levels and correlate them with short- and long-term patterns of precipitation.

❧*ADDITIONAL READING*❧

Andrews, William A. *Freshwater Ecology*. Englewood Cliffs, N. J.: Prentice-Hall, Inc., 1972.

Coker, Robert E. *Streams, Lakes, Ponds*. New York: Harper and Row, Publishers, 1968.

Credland, Peter, and Gillian Stranding. *The Living Waters: Life in Lakes, Rivers and Seas*. New York: Doubleday and Co., Inc., 1976.

Goldman, Charles, R., and Alexander J. Horne. *Limnology*. New York: McGraw Hill Book Co., 1983.

Gravé, Eric V. *Discover the Invisible: A Naturalist's Guide to Using the Microscope*. Englewood Cliffs, N. J.: Prentice-Hall, Inc., 1984.

Hutchinson, G. E. *A Treatise on Limnology*, 1–3. New York: John Wiley and Sons, 1957.

Smith, R. L. *Ecology and Field Biology*. New York: Harper and Row, Publishers, 1966.

PART THREE

THE FLOWING WATERS

Lotic Ecology

CHAPTER FIVE

FROM SOURCE TO SEA

"Look," cried a young voice. The twelve of us wheeled, then splashed, slipped, and mucked along as fast as green algae-covered rocks would allow. Soon we stood frozen in a circle, mouths agape and eyes wide. There, on a gently swaying reed growing in a shallow pool, a homely insect nymph began its miraculous transformation into a graceful, delicate damselfly.

Clinging head up, with its three pairs of legs clutching the reed, the nymph at first appeared to be motionless. Moving almost imperceptibly, it wriggled its head and abdomen out of a split in the skin along its back. I wondered how we must have looked through the glaring compound eyes on each side of the nymph's head, each with ten thousand to twenty thousand facets. Gripping the stalk firmly, it extracted the needlelike, brilliant metallic blue abdomen from the spent nymphal case. Moist, folded wings were held over its back. One hour later, as our group was leaving Tobyhanna Creek and its creatures, these wings were still only half extended.*

This cryptic damselfly nymph emerged on the banks of the Tobyhanna Creek near Pocono Lake in the Pocono Mountains of Pennsylvania. Amid

*Excerpted from "Of Streams and Insect Life" by Michael J. Caduto from Vermont Natural History, November 1982 and Audubon Society of Rhode Island Report, September 1983. Used by permission.

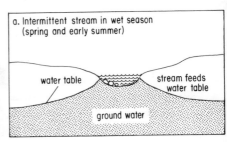

a. Intermittent stream in wet season
(spring and early summer)

water table

stream feeds
water table

ground water

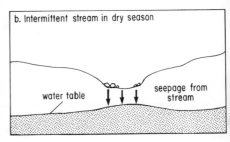

b. Intermittent stream in dry season

water table

seepage from
stream

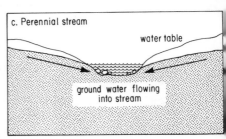

c. Perennial stream

water table

ground water flowing
into stream

FIGURE 5–1: (*left*) This intermittent stream lies dry for most of the summer months, a time when it flows only after a heavy rain. (*Photo by Michael J. Caduto.*) (*right*) Relationship between water table and two types of streams: intermittent and perennial.

the swirling eddies and limpid pools of every healthy stream and river is a host of freshwater plants and animals. We may swim, fish, or canoe in these roily waters, but what do we know of conditions for life along the rocky streams and placid rivers? This chapter looks at lotic ecology, the study of flowing freshwater environments and the life found within.

Flowing waters can range from the tiniest trickle emanating from a cliffside spring to the rampaging waters of Niagara Falls. Some streams are ever-changing environments. *Ephemeral streams* dry up between heavy rainfalls, and *intermittent streams* flow only during the wet seasons of winter, spring, and early summer when the water table is high, and after heavy rains during the rest of the year. While some springs, such as Silver Spring in Florida, are among the most stable freshwater homes, having consistent temperatures and oxygen levels throughout the year, most streams and rivers are dynamic places—a mosaic of many different habitats.

From the clean mountain streams that gradually carry eroded soil and rock from the hillsides, to the broad, lazy meanders that deposit that sediment along lowland riverbanks or carry it out to sea, the flowing waters are sculptors of great power. And yet rivers and streams cover less than one thousandth of the earth's land surface and contain only .005 percent of the world's total supply of liquid fresh water at any given time.

❧THE FLOWING WATER: A COMPLEX HOME❧

What would it be like to live in a stream or river? Challenging! Being dense and viscous, flowing water presents tremendous force. Moving around in the current is difficult at best. With one wrong step, you might get dislodged and battered against rocks and sticks. Ice can cut and grind as it breaks up and rushes downstream. The normal functions of life—eating, breathing, moving or holding fast, laying and hatching eggs—all require special adaptations. You might get covered with sediment after a heavy rain, washed away during a flood, or even stranded on a dried-up streambed.

The current can also be a blessing. In streams and small rivers, as water flows over a rough bottom it becomes a giant mixer, saturating the water with air. Water, being an almost universal solvent, bathes you in a diluted soup of vital, dissolved nutrients, and then flushes your wastes downstream. Most of these nutrients come from leaves that have fallen into the rushing water and soil that eroded from the banks. These nutrients will be used many times as they move downstream; they will be incorporated into the plants and animals along the way.

There is much more to learn about your home in the stream. Each set of conditions outlined below interacts with the other parameters of life in the flowing water to create your total environment. These descriptions compare the upper stream environment with river habitats. Remember that the transition between these two environments is gradual, and that many stretches of flowing water are transitional, possessing characteristics of both stream and river.

DISSOLVED OXYGEN

Dissolved oxygen levels in small streams are usually at the saturation level for water at that temperature, due to the mixing action of the fast-flowing

FIGURE 5–2: The fall leaf drop is a major source of nutrients for the stream. (*Photo by Michael J. Caduto.*)

FIGURE 5–3A: A dam that drains from the bottom. (*Photo by Michael J. Caduto.*)

water. During the heavy fall leaf drop in temperate regions there is much decomposition and respiration, with a corresponding dip in the oxygen level. Daily changes in DO levels are slight, even under cover of ice. Large, slow-flowing rivers experience a daily cycle of dissolved oxygen and carbon dioxide similar to that of lakes, as described in Chapter 2. Oxygen levels are often below saturation, especially when there is a high input of organic matter from plant growth and pollution. A coating of ice causes less mixing of the upper layers with the air and decreases DO levels.

TEMPERATURE

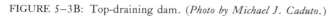

Heavily shaded banks and cool springs help to moderate the temperature of stream water. The churning water tends to remain cool from top to bottom, a crucial factor, since colder water can hold more dissolved oxygen for aquatic life. The warm season temperature can range from 50–80°F (10–27°C), and even higher in rivers. Water is generally warmer in the lower stretches of rivers, with slight thermal stratification occurring in slow, deep sections. Dams can cause the river below to be either cool or warm, depending on whether the river flows from the bottom or over the dam.

FIGURE 5–3B: Top-draining dam. (*Photo by Michael J. Caduto.*)

Some nutrients are recycled within the stream or river, and some are introduced from the surrounding environment. The major producers in a stream are the *periphyton*—attached algae. Examples include green and blue-green algae, as well as pennate and centric diatoms. Periphyton is sometimes referred to as *Aufwuchs*. Streams consist of a series of riffles, runs, and pools, and most primary production occurs among the periphyton on the rocks of the riffles, especially in sunny areas. Dead plants and animals wash downstream and settle in the still pools, where they are decomposed, producing the carbon dioxide and releasing nutrients for further production by plants downstream. This cycle of production and decomposition can be seen as a kind of *nutrient spiral*. The fall leaf drop causes a great input of nutrients and is a major food source for small mountain streams. In many sections of stream the benthic invertebrate animals, such as insects, crustaceans, and mollusks, comprise the major volume of living matter or *biomass*.

Rivers are fed nutrients from the decaying remains of algae produced upstream, by phytoplankton living in the sunny upper layers of the river channel, and by phytoplankton and macrophytes in backwaters and associated wetlands. Additional nutrients are washed into the river channel during periods of high water when surrounding swamps, marshes, and meadows are flooded.

Phosphates (a nutrient) enter the lotic waters as dissolved phosphate and on the particles of eroded soil, as do nitrates, which are also supplied by rainwater. Ammonia is produced by nitrogen-fixing bacteria, which is later

FIGURE 5–4: Nutrient spiral in a stream. Production occurs primarily in the riffles and runs, while major decomposition takes place in the still pools. Nutrients are used for production in the fast water, then the resulting organic matter settles and decomposes in pools, where nutrients are once more made available for production.

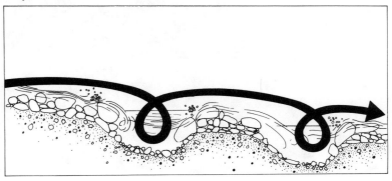

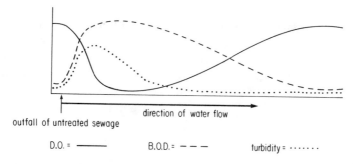

direction of water flow

outfall of untreated sewage

D.O. = ———   B.O.D. = — — —   turbidity = · · · · · · ·

FIGURE 5–5: The effects of severe organic pollution (such as untreated sewage) on the levels of dissolved oxygen (DO), biological oxygen demand (BOD), and turbidity, in a flowing-water ecosystem. Dissolved oxygen levels can drop to, or near, anaerobic conditions just below the outfall. Here, bacteria use up oxygen as they decompose the organic matter from the sewage, as well as the remains of algae that bloom in response to the overly fertile waters. As the organic matter is consumed moving downstream, DO increases as decomposition and respiration drop, photosynthesis rises, and oxygen mixes in from the atmosphere. Biological oxygen demand and turbidity both increase below the outfall, where the sediment load and biological activity are high, and then decrease as the sewage is decomposed and the water gradually recovers from the pollution. Turbidity also contributes to low oxygen levels by decreasing the amount of sunlight reaching the photosynthetic organisms.

converted to nitrate by other bacteria. In hard waters, where there are high levels of photosynthesis and abundant supplies of calcium, calcium carbonate ($CaCO_3$) deposits—called *marl* or *travertine*—may form a white coating over the stream bottom.

Human influence on nutrient levels and the lotic community. Running waters experience a different dynamic from ponds and lakes when excess nutrients and organic matter are introduced. Normally, a balance exists between the action of the bacterial and algal slime encrusting rocks and gravel. Algae produce oxygen, and bacteria use it for decomposition of algal and other organic remains. What happens if this balance is disrupted? Suppose that there is a major point source of nutrients into a river, such as effluent from a wastewater treatment plant. Downstream from the source of nutrients, algae will bloom profusely, feeding on the nutrients in these overfertilized waters. Then, as the algae die, the enormous volume of their remains begins to decompose, and the resulting high rate of respiration—taking in oxygen and giving off carbon dioxide—creates an oxygen sag below the sewage outflow. Those organisms that can tolerate extremely low levels of DO will be found here, often in large numbers: *Sphaerotilus*, a sewage bacterium (often called the sewage fungus) that grows in long, gray filaments; protozoans, including certain species of *Vorticella; Tubifex* worms; bloodworms (*Chironomid* midge larvae); moth fly larvae; and rat-tailed maggots, the larvae of hover flies or flower flies. Gradually, as the water flows downstream, the intensity of oxygen consumption and respiration by decomposers decreases, the nutrients are diluted, and algae and other plants can grow once again. A higher diversity of life is found here where many species of organisms can survive: burrowing mayflies like *Hexagenia* and *Ephemera*, with their plumose gills; numerous larvae of true flies; caddisfly larvae; scuds; leeches; *Planaria*; and water sow bugs are common.

With continued dilution and biological recovery moving downstream from the source of pollution, the river will gradually return to its original conditions above the pollution inflow. Slower, smaller, pondlike rivers often require longer distances to process organic pollution, while faster, well-aerated water will recover more quickly. This type of pollution, if it persists near the mouth of a river, can mask the natural organic substances that salmon use as clues to locate their home stream. This confuses the salmon and hinders spawning runs.

Wastewater treatment plants try to duplicate the natural cleansing process of aquatic ecosystems. First, debris is sieved out, such as sticks, stones, and litter. Primary treatment removes large organic solids with a grid and settling tanks. Under secondary treatment, decomposers are introduced to remove finer organic particles, the water is aerated to provide oxygen for the decomposers, and then the water is further cleared in settling or clarification tanks and chlorinated before it is discharged into the river. This is the most common form of municipal wastewater treatment. Secondary treatment may also use a process that approximates the action of algae and bacteria on the stream bottom. Effluent is sprayed on a trickling filter bed, where it gravitates through slime-coated rocks, gravel, and finally sand, which digests the organic matter in the water and filters out particles. Tertiary treatment can bring wastewater close to drinking standards. It is an expensive method that uses chemicals to remove phosphates, nitrates, and other compounds.

ACIDITY

Acidity is affected in lotic waters by many of the same factors, and with similar affects, as described earlier in general terms (Chapter 2) and in lentic waters (Chapter 3). In addition, acid rainwater can increase the acidity of streams, while surface runoff, which may have been buffered by the surrounding soils, can moderate the pH of stream waters.

WATER LEVEL FLUCTUATION

In ephemeral and intermittent streams, the streambed is usually located above the water table. These streams recharge the ground water and dry up when runoff is insufficient, or when water tables drop during dry periods. *Perennial streams*, being situated at or below the water table, are fed by ground water throughout the year.

WATER MOVEMENT

Laminar flow is the smooth movement of water through unobstructed stretches of the stream. Objects in the stream and friction between the water and the bottom disrupt the flow and create turbulence. *Eddies* are a special kind of contrary turbulence that form circular upstream currents behind rocks and other objects, and along the edges of a stream channel. The next time you are canoeing a river past a large rock, swing the bow around behind the rock so that the canoe is facing into the current. If the eddy is strong enough you can take your paddles out of the water and the current will hold you upstream against the rock!

FIGURE 5–6: Whirligig beetles take refuge from the current in this eddy behind a rock. (*Photo by Michael J. Caduto.*)

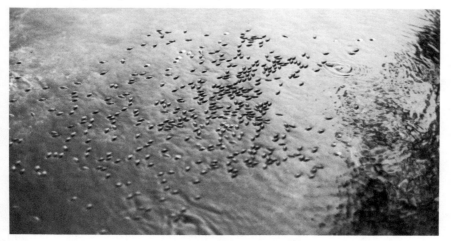

DISCHARGE

The discharge of a stream or river is the volume of water passing a point in a certain amount of time. It is affected by the amount of land area feeding the stream and its slope, the permeability and water storage capacity of the surrounding soils, and the rainfall pattern. *Current* or *velocity* measures the distance traveled by the water during a certain length of time. Velocity depends on the depth of the stream or river, the slope, and friction along the stream due to the texture of the bottom and shape of the channel. The highest measurements of velocity and discharge occur during the spring thaw and rainy season. Velocity is highest just under the surface because friction between the water and air slows the surface water slightly. Faster currents are found at the outside of a bend. This force erodes the outer edge of the channel, while the slow water on the inside of a turn is often a site of sediment deposition. Average velocity is usually at about two thirds of the stream depth. Although bubbling streams appear and sound as if they are running faster than a quiet river, the velocity may actually be greater in the lower stretches of a river system. *Floods* or *spates* can devastate the life along a stream bottom by washing the sediments downstream, along with the plants and animals living on and in the bottom, which are often crushed in the rolling rocks and gravel or dashed against objects by the current.

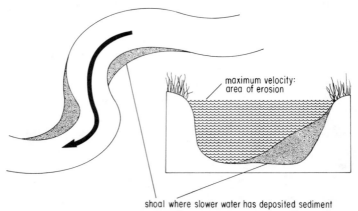

maximum velocity: area of erosion

shoal where slower water has deposited sediment

FIGURE 5–7: Cross section of a stream at a sharp bend.

BOTTOM TYPE

Closely related to the velocity and discharge of a stream or river is bottom texture. Fast waters, because they contain more energy, can carry away all but the largest particles of soil, sand, or rock. So faster waters have a coarse bottom of bedrock, rubble, or gravel. These are generally found in the upper stretches of a river system. As it enters the lowlands, water slows and drops finer sand, silt, and clay. Gravel and rubble are productive environments to live in because they are fairly stable and allow aerated water to enter, which brings in food and flushes wastes away.

THE EFFECTS OF SURROUNDING HABITATS

Riparian or riverside vegetation and soil type are crucial environmental determinants of the flowing waters. Plant cover along the banks shades the water and moderates temperature, the roots hold soil in place to prevent erosion, and nutrients are added to the water by falling leaves and sticks. Soils over hard rocks tend to be less fertile and more acid, containing few

FIGURE 5–8: The types of stream and river bottoms, listed in order from those found in faster to slower water: A. bedrock; B. rubble; C. gravel; D. sand; E. silt and mud. (*Photos by Michael J. Caduto.*)

A

B

C

D

E

THE STREAM AND RIVER AT A GLANCE

Fast Streams

- Dissolved oxygen at saturation levels

- Often cooler, temperature uniform top to bottom
- Production within the stream occurs mostly in riffles, with a large input of organic matter from outside the stream coming from the leaf drop each fall and other outside sources

- Water level may show relatively high fluctuation
- Riffles, runs, pools in small V-shaped valleys

- Shallow
- Steep slope with bottom of gravel, rock
- More often clear water
- Often shaded by overhanging vegetation
- Erosion occurring over most of its length

- Narrow channel
- More diversity of organisms in any given location in the channel

Large, Slow Rivers

- Dissolved oxygen high, but follows a lakelike pattern
- Warmer, sometimes thermally stratified
- Some production in the river may come from river phytoplankton in main channel and still waters along the banks; nutrients are introduced from upstream areas, soil erosion, and surrounding wetlands and sloughs
- Water level more stable

- Generally fast, even flow through broad, flatter lands with rounded banks, U-shaped valleys
- Deep
- Gradual slope with bottom of mud, silt, sand, or clay
- Turbid
- Unshaded toward the middle of the channel
- Area of erosion in faster waters and deposition in slow water, along floodplains and at the sea
- Broad course
- Greater diversity of organisms when measured along length of the channel

leachable nutrients and providing smaller volumes of plant material to the stream. Soft waters are found here. Hard, fertile waters and relatively lush forests tend to be found over rocks that are more easily dissolved. This type of streamside environment contributes a relative abundance of nutrients to the stream.

SUMMARY OF THE CONDITIONS IN STREAMS AND RIVERS

The patchwork quilt of different conditions that is home for the plants and animals of flowing waters is largely determined by changes in discharge into the stream or river, the input of soils from the surrounding area, and the kind of vegetation growing there, all of which are in turn influenced by the local climate. Some remarkable adaptations exist to the extreme conditions of ephemeral or intermittent streams. For instance, some bugs and beetles can fly away when the stream dries up; some water fleas form drought-resistant resting stages. Other insects pupate in the mud or debris, timing their life cycle to be in a quiescent stage during the drought. Death comes to those that cannot adapt.

❧ MICROHABITATS ❧

Within the changing worlds of the stream and river, each organism finds a place that meets its survival needs. To understand what this means, we need to think small—sometimes very small!

A rock in the current may look like just that, until discerning eyes, aided by a hand lens, can ferret out the rock's secrets. On top there is a coating of slippery encrusting algae thriving in the sunlight. Small black dots, on closer examination, turn out to be hundreds of tiny black fly larvae swaying in the current. The upstream side of the rock is swept clean by the rushing water. Facing downstream, there is a small eddy that offers protection for a snail that grazes on algae. And in the dark shadow beneath the rock are found some nocturnal insects that have taken cover from the daylight: several mayflies and stoneflies. These insects have a flattened shape that enables them to cling closely to the rock. Here there is a space, ranging from less than .04 to .1 inch (less than 1 to 3 millimeters), called the *boundary layer*, where the friction between rock and water slows the current significantly. (See Fig. 5–9, p. 154) This layer becomes thicker when the water is cooler and thus more viscous. Even so, there is a *critical velocity* of water for each species above which the insects would be washed downstream.

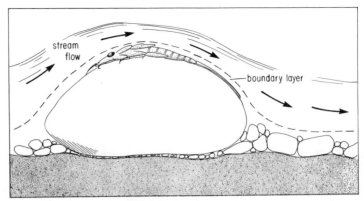

FIGURE 5–9: Mayfly nymph in boundary layer. Friction between the water and the stream or river bottom causes the boundary layer, a thin film of slow water where insects and other organisms are protected from being washed downstream.

Along the edge of the stream are two other important habitats. Cushions of fountain moss, which provides shelter and food, harbor many microscopic organisms and larger animals. Patches of leaves that have become caught between rocks, or a rotting stick near the shore, are also important debris habitats inhabited by many small animals.

ᔕLIFE IN THE FLOWING WATERSᔕ

Rocks, macrophytes, and debris provide some of the most important habitats for stream organisms. The algae and moss growing here are food for the grazers, such as snails, which will also eat fungi, bacteria, and rotifers. The decomposers—like many bacteria and the fungus *Hyphomycetes*—break down plant remains. In this way these nutrients become more available for plant growth. *Shredders* are organisms that eat larger plant remains. The shredders include the case-building caddisflies, crane fly larvae, many mayfly nymphs, and some herbivorous stoneflies.

Other benthic invertebrates are drift feeders, filtering out small plants, animals, and debris from the running water. *Drift* is an integral part of the stream, consisting of algae, bacteria, detritus, and invertebrates that are washed downstream by the current. Many invertebrates, insects in particular, are voluntary components of the drift, releasing themselves into the current at dusk and nighttime hours. Predators, such as trout, feed on other animals in the stream, including members of the drift community. Some lotic plants, and many animals, are active even during the winter, although growth and activity slow down during the colder months.

The remainder of this chapter will look closely at the plants and animals of three major lotic habitats: springs, streams, and rivers.

❧SPRINGS❧

If you began your study by seeking the source of a river, in many cases you would eventually find a series of springs seeping from a wooded hillside. Springs are places of clear, clean, and cool water that has just emerged from the ground-water stores. The average temperature of springs in temperate areas is around 46.4°F (8°C).[1] Watercress is common here, along with some invertebrates such as aquatic sowbugs, black fly and caddisfly larvae, some beetles, snails, and salamanders. Springs are commonly used as watering holes by minks, raccoons, and other larger animals.

Where the flow from many springs melds to form a nascent stream, exposure to the air has since mixed oxygen into the cool water. The shady young stream flows down over stones, rocks, and patches of exposed bedrock that are blanketed in spots with expansive green carpets of fountain moss. Some tiny beetles and mayfly nymphs can be seen in their moss habitat, while a wolf spider waits patiently nearby to catch its dinner. If the bedrock

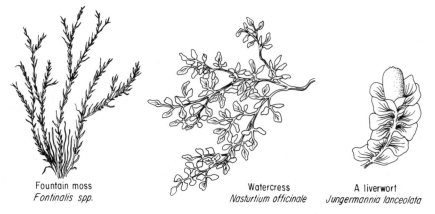

Fountain moss
Fontinalis spp.

Watercress
Nasturtium officinale

A liverwort
Jungermannia lanceolata

FIGURE 5−10: Plants of the cold spring waters. Fountain moss (*Fontinalis spp.*). *Size:* 3 inches (7.6 centimeters). Watercress (*Nasturtium officinale*). *Size:* to 12 inches (30.5 centimeters). A liverwort (*Jungermannia lanceolata*). *Size:* .5 inch (1.3 centimeters).

is smooth and steep enough, you can shoot down this soft, living slide into a pool of water below. The blue-green alga *Rivularia* makes the rocks slippery and difficult to stand on. Watercress grows submerged in here. Pull a small piece of leaf from this aquatic mustard and experience its hot, peppery flavor. If you are lucky, you may find the beautiful, tiny fronds of fragile fern growing on the surrounding moist rocks. Liverworts, such as *Jungermannia lanceolata*, are common on the cool rocks of the upper stream.

Hot springs and underground waters are notable freshwater habitats in close association with surface springs. Some small insects, especially species of midge larvae, have been found in hot springs at temperatures up to 122°F (50°C). Certain roundworms can live in water as high as 140°F (60°C). And a species of blue-green algae has been found living in hot springs at 194°F (90°C) in Yellowstone National Park, Wyoming! The dark environments of caves, wells, and other subterranean freshwater environments cannot support the growth of green plants; the nutrients for these habitats enter from the surface. Fish, amphibians, and insects are blind and light-colored, having no need for sight or pigments to protect their tissues from the sun's ultraviolet rays. Flatworms and crustaceans are also found here.

❧ *STREAM LIFE* ❧

Moving downstream, you pass numerous small, steep-sided valleys, each with its own streamlet joining into the now bubbling stream. As your stream gurgles and gropes its way down over rocks and gravel, it presents to plants and animals a suitable environment of highly oxygenated water and a somewhat stable bottom replete with a variety of habitats in which to live.

STREAM PLANTS

Many of the forms of algae that grow in the stream have been mentioned: centric and pennate diatoms, green and blue-green algae. Some red and golden algae are found here as well. The growth and occurrence of different algae is determined by current speed, DO levels, pH, available nutrients, water temperature, and the amount of sunlight present. Plant populations are also affected by the grazing and munching of herbivores. Different species of algae will grow on rocks, logs, other plants, or on the bottom. In general, algae require water that is not too acid, being of moderate pH.

The adaptations of algae and moss include flattened and tightly clumped growth habits to direct water over and around the plant mass, and a

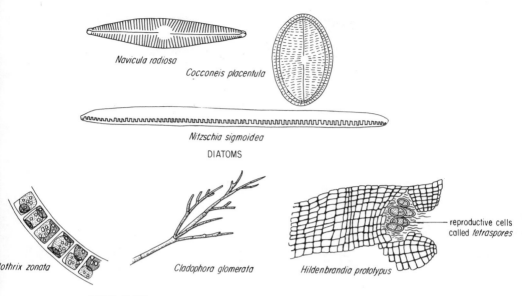

Navicula radiosa

Cocconeis placentula

Nitzschia sigmoidea

DIATOMS

Ulothrix zonata

Cladophora glomerata

Hildenbrandia prototypus

reproductive cells called *tetraspores*

GREEN ALGAE

RED ALGA

FIGURE 5–11: Algae of the stream. *Size:* Microscopic detail shown.

sturdy and spreading attachment to the substrate. Some stream algae produce new generations very quickly to replace those that are constantly being washed away by the current and eaten by herbivores. Diatoms are the most well adapted to cold water; although they are most numerous during the springtime, some are also present in winter.

The larger plants of the streambed—fountain moss, water hypnum, liverworts, and watercress, among others—are also highly adapted to their flowing world. Stems tend to be short and flexible, and leaves are small to produce less drag in the current. Most large plants die back to roots or produce resting stages during the winter. Two common riparian plants are sedges and coltsfoot.

Stream Animals

What would you need to live underwater? What changes would you have to make to live amid the effervescent rapids or in a quiet pool? The necessary adaptations would render you unrecognizable.

Stream animals have adapted well to their changeable environment. Their main food-getting techniques are grazing, filter feeding, and predation. Filter feeders are prevalent because the current carries food along its course, in a sense, delivering groceries right to the front door! Keep in mind the importance of the fall leaf drop, the major source of energy coming from outside the stream each year. This larder supports growth and activity among stream animals in the fall and early winter.

Many stream insects and other animals look virtually the same the world over. This is thought to be the result of highly successful dispersal mechanisms among some cosmopolitan species. *Convergent evolution*, the separate development of similar adaptations due to life in like environments, has also played a role that is especially obvious among the stream insects. Many North American species are hard to tell apart from those found in the Rhine of Western Europe, the Hwang Ho or Yellow River of northeastern China, and the Congo in Africa.

On a walk to a nearby stream, in the still pools at the water's edge, you may find two familiar creatures flitting along the surface.

Gerris, the water strider, also called the pond skater and Jesus bug, is known by almost everyone. The feathery tips of the water strider's legs skate along on the surface as it prowls for unlucky insects that have fallen in. Tiny claws that are set back on the front legs to avoid breaking the surface film are used to handle captured insects. The dimples in the surface film made by the strider's legs cast a shadow on the bottom that looks like a piece of popcorn. These insects use the water's surface much as a spider uses its web to catch prey, locating their food by sensing vibrations in the surface film. Once a victim is captured, the strider uses piercing-sucking mouthparts to finish off the meal. The next time you watch some water striders, catch one of the many mosquitos that are biting you and toss it into the stream. Then watch the striders' feast begin.

FIGURE 5–12: Water strider (*Gerris spp.*). *Size:* .8 inch (2 centimeters). Ripple bug (*Rhagovelia spp.*). *Size:* .2 inch (5.1 millimeters).

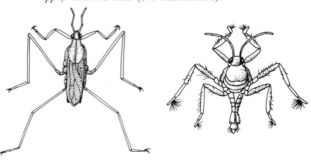

Water strider
Gerris spp.

Ripple bug
Rhagovelia spp.

Another species, the broad-shouldered water strider, frequents the fast waters of the stream in search of food washed down by the current. So named for the fact that its body is broadest in the middle, this ripple bug is a brownish or black color.

In the bend of a lazy meander you may encounter another common denizen of the stream's surface—a group of frantic, never-tiring whirligig beetles.

As you crouch along the stream bank, shift your stance and your gaze to look beneath the surface, beyond the creatures whose movements interrupt the reflection of forest and sky. There, crouching over the gravel and rocks, you may be lucky enough to see a riffle beetle covered with a shiny coating of air clinging to the pile of fine hair that surrounds its body. This beetle belongs to another group of stream insects, those that spend most of their lives submerged and free-swimming, yet need to surface occasionally for air. The larva of the riffle beetle is the well-known water penny, which has a flat, oval shape that is held close to the sides of stones. Adults lay eggs that hatch into larvae, which is the only life stage to overwinter. During the next warm season the larvae will pupate under rocks or sticks at the water's edge.

Yet there are only so many stream creatures that can be seen from the bank. A pair of old sneakers, a hand lens, and a wire collecting screen are all you will need to explore the mysteries of the stream's bottom. Some specific collecting techniques are described at the end of this chapter.

Truly aquatic animals spend entire periods of their lives submerged. They must be sought amid the rocks, gravel, and plants found under the roiling water. Since many of these creatures are nocturnal, they remain well

FIGURE 5–13: Riffle beetle (*Psephenus herricki*). *Size:* .2 inch (5.1 millimeters). Water penny: riffle beetle larva. *Size:* .3 inch (7.6 millimeters).

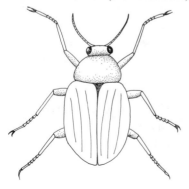

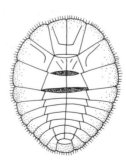

Riffle beetle
Psephenus herricki

Water penny: riffle beetle larva

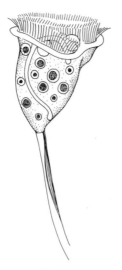

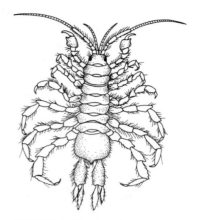

FIGURE 5–14: The protozoan
(*Vorticella spp.*). *Size:* Microscopic
detail shown.

FIGURE 5–15: Aquatic sowbug.
Size: .8 inch (2 centimeters).

hidden during the day. One kind of stream protozoan is *Vorticella*, which catches its food of small animals, algae, and detritus with an entrapping cup. It is attached to the bottom by a long, thin stalk.

Streams are not without their worms and wormlike animals. Waterfalls are a favorite haunt of the net-winged midge larvae. How does one hold onto rocks in a waterfall? Try it sometime! You will see why these larvae have a row of true suction cups on their ventral surface, as well as silk threads and sticky saliva—a perfect set of tools that enable them to move like an inchworm along the rocks, alternately releasing and gripping with their front and rear suckers. Some common worms of the stream include nematodes (roundworms), *Planarians* (flatworms), leeches, and the bristleworm, *Nais*. The last two are relatives of the earthworm.

Almost every aquatic environment has its representatives of the Crustacea. Lotic species of scuds and crayfish are found here. If you see something creeping slowly away as you turn over a rock, and it looks like the pillbugs or sowbugs that hide under logs and in damp basements on land, it is an *isopod*, the aquatic sowbug. Related to its familiar terrestrial cousin, the aquatic sowbug has seven pairs of legs and is flattened when viewed from the side. It breathes with gills while foraging plants, detritus, and dying animals. After mating, the larger female lays up to 250 eggs and may raise two broods each year. Like most of the crustaceans, it has two body parts: a fused head and thorax, called the *cephalothorax*, and an abdomen. Most are gray but they can range in color from yellow to black.

As you lift another rock nearby, some tiny (less than .2 inch or 5 millimeters) black fly larvae may be seen swaying in the current. They are small enough to be protected by the slow water of the boundary layer. There are approximately ninety species of black fly or buffalo gnat in the United States. The larvae are unmistakable, with their bulbous butts anchored firmly in place with silk threads, sticky saliva, and over 100 hooks set in a radial pattern. Sievelike hairs project from the sides of the head to strain algae, tiny animals, and plant debris from the water. The larvae are sometimes so abundant that they form a dark carpet over the rocks. If dislodged, they will let out a silk safety line and creep back up the current. One to two broods of eggs are laid underwater each year on logs, rocks, and plants of very swift, shallow waters. In two to six weeks the larva pupates and, when mature, the adult will emerge, float up on a bubble to the surface, and then fly away. The adults of some species of black fly are infamous for their incessant biting and hovering about the eyes.

Crane flies and soldier flies are also true fly larvae of the streambed. Soldier fly larvae eat algae, detritus, and small animals. Adult crane flies are well known as "mosquito hawks," often seen hovering in the shady dark hollow of a tree or fluttering on screen doors at night. The aquatic larvae are cream-colored and maggotlike in appearance, over 1.5–2 inches (3.8–5 centimeters) long, and bear some appendages on the tail that look like the nose of a star-nosed mole. The two dark spots near these protrusions are

FIGURE 5–16: Black fly larva (*Simulium venustum*). Note the sievelike, food-gathering structures near the larva's mouth (top side). *Size:* .2 inch (5.1 millimeters). Black fly adult. *Size:* .2 inch (5.1 millimeters).

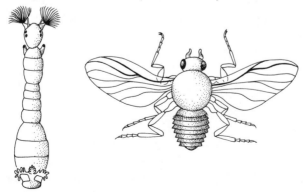

Black fly larva
Simulium venustum

Black fly adult

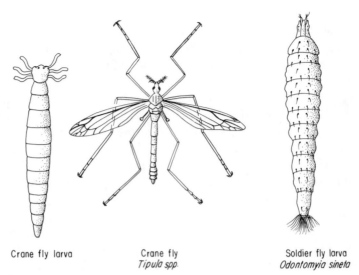

Crane fly larva   Crane fly   Soldier fly larva
Tipula spp.   *Odontomyia sineta*

FIGURE 5–17: Crane fly larva. The dark spots between the six tail appendages are spiracles, which aid in getting oxygen. *Size:* 1.5–2 inches (3.8–5 centimeters). Crane fly (*Tipula spp.*). *Size:* .8 inch (2 centimeters). Soldier fly larva (*Odontomyia sineta*). *Size:* .6 inch (1.5 centimeters).

spiracles, openings that can be used like a snorkel to breathe in shallow water. Some larvae are predaceous, while others are scavengers.

    Turning over some more rocks, you are likely to find some common stonefly or mayfly nymphs, which are not true flies. You may note that the names of true flies are written in this book as two separate words, while other insects with "fly" in their name are written as only one word. Flattened, with strong claws on their feet (stoneflies have two claws on each foot, while mayflies have only one), they dwell in the .04–.1 inch (1–3 mm) boundary layer. Mayflies have gills on their abdomen and usually three, sometimes two, tail appendages called *cerci*, which are thought to be used in reproduc-

FIGURE 5–18: Stream mayfly. *Size:* .6 inch (1.5 centimeters).

tion by adults. Stoneflies have thoracic gills and only two cerci. Depending on the species, stonefly nymphs spend one to two years foraging along the bottom, while mayflies may have one or more generations per year. Virtually all mayflies and some stoneflies eat plants or detritus, while some stoneflies love to dine on delicate mayfly nymphs and other insects during their nighttime forays.

Stoneflies are common in good trout streams, wherever the water is clean, cool, and swift flowing. In fact, they are a favorite meal for trout. Many species take two growing seasons to mature, and adults can usually be found from November through August. Some species of the herbivorous giant stoneflies of the family *Pteronarcidae* grow from egg to adult in 10–11 months, being born one summer and maturing the next. Giant stoneflies can eat 29 percent of their weight each day. Tiny (.5 inch or 12 millimeters) winter stonefly adults can be found emerging from the stream on sunny days from January through April. When a nymph is ready to moult it crawls onto a rock and the adult emerges from a split on its back. Finding one of these

FIGURE 5–19: Giant stonefly. *Size:* to 1.7 inches (4.3 centimeters). (*Photo by Michael J. Caduto.*)

insects crawling on the snow looking for plants to eat on a 14°F (−10°C) winter day is truly inspiring. In general, stoneflies that emerge during the summer are nocturnal and do not eat in their short-lived adulthood, which usually lasts a few weeks. After mating, depending on the species, eggs are laid either over the water, on the surface, or submerged.

Living as part of the stream benthos for one year or less, a mayfly nymph eventually hatches into an ephemeral subadult called a *dun* or *sub-*

imago, and then moults into an adult whose sole purpose is to reproduce. Adults hold their wings together vertically over their backs. In fact, most adult mayflies have reduced or nonexistent mouthparts and guts—they do not eat. They mate in flight during a frenzied one or two days, after which the male dies. The female often perishes while trying to reenter the water to lay her eggs on the bottom. Burrowing mayflies live in silty or clay-bottom streams. They are famous for dramatic hatches that produce giant swarms. In some towns along the Mississippi River, dead and dying insects and cast subimago skins literally form a layer several inches deep on village streets. Mayflies either burrow, sprawl on the bottom, or dwell on rocks or in the open water.

While they are more common in lentic environments, there are some stream-loving damselflies. The graceful black-winged damselfly is a frequent sight in June and July. They may be found perched on a low tree or shrub, waiting in ambush until an unsuspecting prey flies by. Females have a white dot or *stigma* near the tip of each wing on the leading edge. The name of these delicate creatures comes from the French *demoiselles*, meaning "damsels."

FIGURE 5–20: ♀ Black-winged damselfly (*Calopteryx maculata*). *Size:* body, 1.9 inches (4.8 centimeters).

Most of these insects, except for the true bugs and beetles, are harmless to hold, but be careful of the powerful jaws of the hellgrammite, the dobsonfly larva! These insectivorous larvae live 2–3 years, can grow to be 3 inches (7.6 centimeters) long, and are avidly sought for bait by trout fishermen. Their abdomen is covered with numerous gill filaments; they can also breathe for a long time out of water using their spiracles. When ready to pupate, the larva climbs onto the stream bank and burrows under a rock up to 164 feet (50 meters) away. Encountering a full-sized male with its 1-inch-long (2.5 centimeters), curved mandibles is nothing short of awesome. Deceptively sinister looking, these claspers are really used to hold the female while mating. A member of the order *Megaloptera*, dobsonflies are closely related to fishflies and alderflies. Dobsonflies and fishflies are nocturnal, while alderflies are commonly seen during the day.

By now, if you have been persistent and observant in your search among the rocks and gravel, you are likely to have discovered a tube-shaped caddisfly home of leaf pieces or sand grains. Of all the insects encountered in

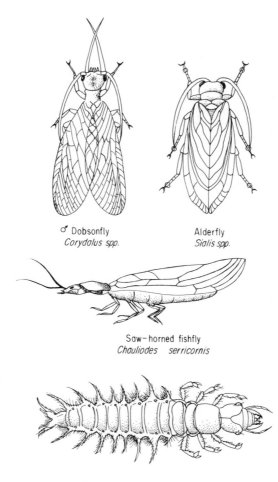

♂ Dobsonfly
Corydalus spp.

Alderfly
Sialis spp.

Saw-horned fishfly
Chauliodes serricornis

Hellgrammite (Dobsonfly larva)

FIGURE 5–21: Representatives of the suborder Megaloptera. ♂ Dobsonfly
(*Corydalus spp.*). *Size:* 2.7 inches (6.9 centimeters). Alderfly
(*Sialis spp.*). *Size:* .6 inch (1.5 centimeters). Saw-horned fishfly
(*Chauliodes serricornis*). *Size:* 2 inches (5 centimeters). Hellgrammite
(Dobsonfly larva). *Size:* to 3 inches (7.6 centimeters).

the lucid pools and rolling ripples of the streambed, none are more curious,
inventive, and inspiring than the omnivorous larvae of caddisflies.

There are three major kinds of caddisflies. Some, like *Rhyacophila*, are
free-swimming as larvae. Case builders are the engineers of the stream,
constructing their homes by weaving an intricate tube of silk threads that is
closed at one end. Depending on the species and the age of the larva (some
larvae use different materials as they get older), either sand grains, leaf

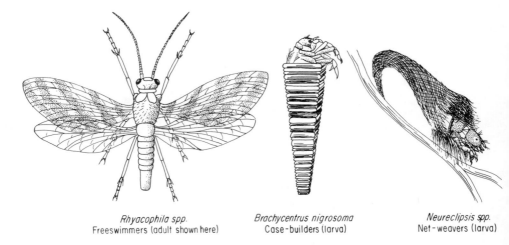

Rhyacophila spp.
Freeswimmers (adult shown here)

Brachycentrus nigrosoma
Case-builders (larva)

Neureclipsis spp.
Net-weavers (larva)

FIGURE 5–22: Representatives of three kinds of caddisflies. The log cabin cases of *Brachycentrus spp.* and the finely woven nets of *Neureclipsis spp.* are true marvels among nature's designs. *Rhyacophila spp.*, larvae are freeswimming (adult shown here). *Size:* to .5 inch (1.3 centimeters). *Brachycentrus nigrosoma*, Case-builders (larva). *Size:* .6 inch (1.5 centimeters). *Neureclipsis spp.*, Net-weavers (larva). *Size:* net to 1.5 inch (3.8 centimeters).

pieces, or small sticks are glued to a silk tube, often in a neat spiral pattern. *Brachycentrus*, the log cabin caddisfly, fashions a square home that lives up to its name. One genus, the *Helicopsyche*, was originally classified as a snail when first discovered because its small case is shaped like that of a snail. The *hydropsychids* are net weavers, catching their food—debris, algae, and small animals—in tiny nets. *Neureclipsis* spins a net that looks like a tiny French horn.

Caddisfly larvae are mostly herbivorous, eating moss, algae, and dead leaves. Abdominal gills are used for breathing, which is aided by the undulating motion of larvae in their cases to help keep fresh water circulating. Two tiny prolegs or hooks help to anchor them in place.

When ready to pupate, caddisflies will attach their homes onto a rock and close themselves in. In around two weeks or longer, during the warmer months, the adult, mothlike caddisflies emerge. There are one or more generations each year. The nocturnal adults have long antennae and large wings that are held over their back like a tent. After only a few days, during which little is eaten, mating occurs, and the female embarks on a perilous egg-laying mission to the stream bottom. A few species lay their eggs at the surface.

Stream insects are true marvels of nature's design. Being of lineage ancient beyond our conception, their simple yet infinitely successful adaptations to stream waters have allowed them to survive through many long periods of traumatic change on earth.

But you have not uncovered all of the living treasures to be found on the streambed. Snails are common here, and a close relative, the cone-shaped limpet, may be seen. Most stream snails have gills and a hard, fingernaillike plate called the *operculum* that closes off the snail's case when the foot is withdrawn.

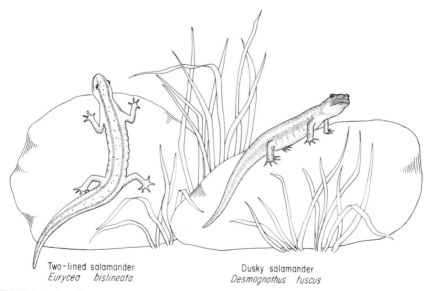

FIGURE 5–23: Limpet (*Lanx patelloides*). *Size:* .4 inch (1 centimeter).

Crayfish, such as some species in the genera *Cambarus* and *Orconectes*, are a common sight among the stream gravels. They exhibit many of the same characteristics as their lentic relatives. Some stream species have been noted to burrow into the bottom during the cold winter season.

Two of the hardest animals to catch are the dusky salamander and the two-lined salamander. They are commonly hidden among the rocks and gravel, coming out at night to feed.

Two-lined salamander
Eurycea bislineata

Dusky salamander
Desmognathus fuscus

FIGURE 5–24: Two common stream salamanders. Two-lined salamander (*Eurycea bislineata*). *Size:* 3 inches (7.6 centimeters). Dusky salamander (*Desmognathus fuscus*). *Size:* 3–5 inches (7.6–12.7 centimeters).

Fish are major predators of the open water. Many feed on detritus, benthos, drift, and insects that fall into the stream from the banks. Some are territorial, protecting their feeding and mating sites and living in one part of the stream for most of their lives.

When people think of streams they think of trout, perhaps more than any other fish. Trout prey mainly on benthic invertebrates and terrestrial insects that have fallen into the water. They love clean, cool water with high levels of dissolved oxygen in which to spawn. Gravelly rivers and pools are important places where they catch food as it washes downstream. The brook trout can be told by the white lines along its pectoral, pelvic, and anal fins, and "worm tracks" running down its back. Very clean waters are needed to sustain healthy numbers of brookies. If you see a rosy lateral stripe and white lines on only the pelvic and anal fins, it is the rainbow trout. Rainbows were first introduced into eastern waters from the West Coast of the United States. The brown trout is strikingly different in appearance. Its combination of black spots with white halos, large red spots along the lateral line, and an orange edge on the adipose fins on its back are good keys. The brown trout is a nonnative fish that was introduced into North America from Europe in the 1800s.

Another salmonid, a member of the trout family, is the Atlantic salmon. Being anadromous, salmon spawn in fresh water but live most of their

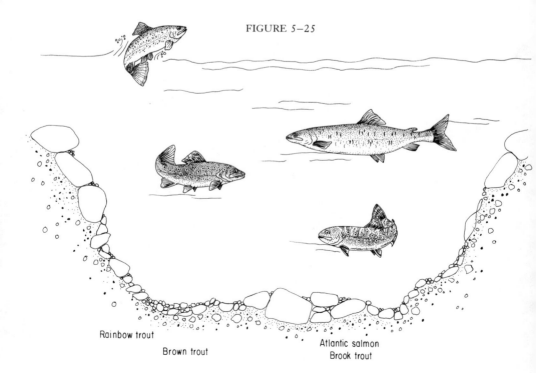

FIGURE 5–25

Rainbow trout

Brown trout

Atlantic salmon

Brook trout

lives in the sea. Good trout streams are ideal spawning grounds for salmon. Young salmon, called *alevins,* hatch here and live for a time off the remnant yolk sac from their egg. After two years the young salmon reaches the smolt stage and is capable of life in the sea. The southwest coast of Greenland is a prime saltwater salmon ground. After spending 1–2 years at sea, the salmon return to their birthplace. It is thought that they locate spawning grounds by the unique odors and chemistry of the water, and they may also use the sun, moon, and stars to aid in navigation. But their homeward voyage of thousands of miles remains largely a mystery. On their way back to their ancestral spawning grounds they use the upward backeddies at the base of waterfalls, and the powerful thrust of their tail, to catapult almost 6.6 feet (2 meters) high in a single leap! Once they have spawned most Atlantic salmon die, their energy reserves exhausted, although some do survive to spawn a second time.

The bottom-feeding sculpin, alias miller's thumb or muddler, with its mottled green colors, is well camouflaged against the algae-encrusted bottom. This comical-looking, slightly pop-eyed fish eats benthic invertebrates and uses its pectoral fins for gripping the bottom.

Numerous other small fish inhabit the stream. American shad spawn here in the springtime. The elaborate spring mating rituals of the three-spined stickleback can sometimes be seen. Although sticklebacks are found

FIGURE 5–26: Minnows of the swift waters.

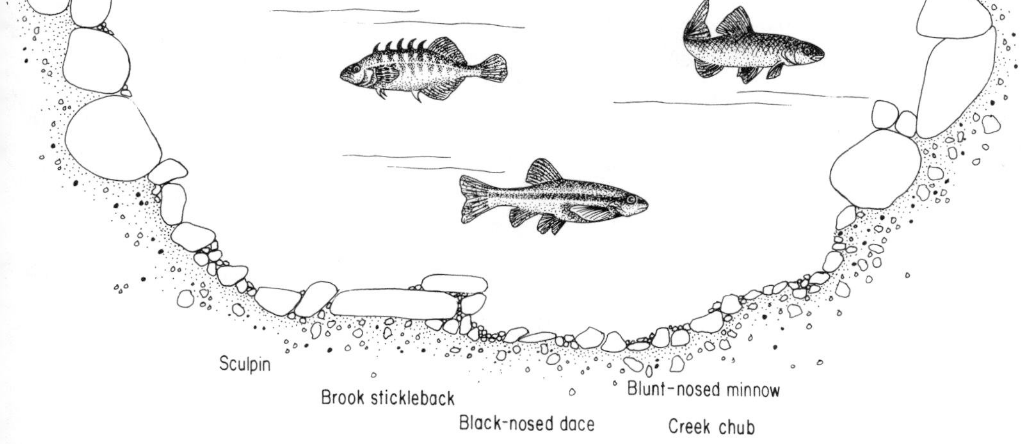

Sculpin

Brook stickleback   Blunt-nosed minnow

Black-nosed dace   Creek chub

primarily in coastal waters, the brook stickleback is a freshwater species. And numerous small minnows (*Cyprinidae*), those unlikely relatives of the much larger carp, frequent these waters.

Amid the moist, shady grottos and waterfalls of the upper stream you may be lucky enough to hear the startlingly loud call of a tiny warbler, the waterthrush. The northern and Louisiana waterthrushes, with yellow and white stripes over their eyes respectively, bob along the mosses and ferns, eating insects.

The flutelike call of the water ouzel is heard around western streams. These amazing birds dive into the stream, walk on the gravels and rocks, and use their wings to "fly" underwater in the rapids. Insects make up much of their diet, including caddisfly larvae and stonefly nymphs. Oven-shaped nests of moss are built on the moss-covered ledges along the stream.

Mammals, of course, frequent the stream: beavers to find food or perhaps to create a home; raccoons, opossums, and water shrews to feed and drink; and black bears to feed at those few streams where the salmon still spawn.

A LEAF FALLS INTO THE STREAM

Before you move downstream to explore the river, consider the simple event of a leaf falling into the stream. Watch what happens to this leaf to see how dynamic the stream really is.

An aspen leaf, a food preferred by many stream insects over other leaves, flutters into the current and lodges in the midst of a clump of other aspen leaves stuck in the crotch of a submerged log. Immediately the nutrients begin to leach from the leaf, and it is colonized by fungus and bacteria. Within one week the leaf weighs less than one-fourth its original weight. But the lightened leaf is now more suitable for insect food, and the bacteria and fungi colonizing the leaf provide proteins and carbohydrates for hungry insects. Black fly larvae appear, as do some grazing species of mayfly: *Ephemerella* and *Stenonema*. By the end of the fourth week very few *Baetis* are found here, but great numbers of *Ephemerella* are present. Barely a skeleton of the original leaf exists when the sixth week arrives, and *Ephemerella* are the only insects remaining. The leaf has become part of the living stream community.

❧LIFE IN THE RIVER❧

Once again, as with lakes, ponds, and marshes, the boundaries between streams and rivers are transitional and somewhat artificially imposed. Along a course of flowing water there are fast and slow stretches alternating between gravel, rubble, sand, and silt bottoms. Moving downstream, however, the proportions of sand, silt, and finally muddy substrates become greater until the river is a wide, deep body of water snaking through rolling hills and coastal plains, then finally forming a delta where the water slows and drops its sediments on entering the ocean.

Stretches of river with bottoms of shifting and drifting sand support the lowest diversity of plants and animals. There is little aeration in the sand, and the constant movement of the substrate buries small animals and provides poor stability for rooted plants, which are easily washed away during a heavy rain.

Some sedges and grass may grow along the shore, and arrowhead, with its deep-rooted tubers, seems to tolerate the changing conditions of the sandy riverbank. When the water level drops in late summer, the seeds of coltsfoot, marsh Saint Johnswort, and grasses will sprout up on small sandbars and islands.

Correspondingly few animals are found here because there is little cover or food. Some species of damselfly, leech, and caddisfly occasion this environment. Mink frogs and leopard frogs may be nearby feeding on the larvae of alderflies and mayflies, while *Planaria* crawl in the shadows looking for a dead animal to feed on.

PLANT LIFE OF THE RIVER

Farther downstream, where the river is broad and deep, the lotic world takes on much of the character of still water habitats. Toward the center of the channel insufficient sunlight penetrates the turbid water to support plant growth on the bottom. Planktonic diatoms, green algae, and blue-green algae are common producers of the open river. Overall, however, phytoplankton are not abundant in the river channel. In the shallows along the bank, and in the backwaters, macrophytes are a source of shelter for animals. Diatoms and other species of algae provide food here. Larger plants, of the

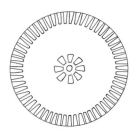

FIGURE 5–27: A river diatom (*Cyclotella stelligera*). *Size:* Microscopic detail shown. (Redrawn with permission from W. T. Edmondson, ed., *Freshwater Biology*. New York: John Wiley and Sons, Inc., 1959, p. 188.)

kind found in marshes and ponds, create shaded areas and microhabitats: willows, tussock sedge, cattails, arrowhead, *Iris*, *Sphagnum* moss, pickerel-weed, and reed canary grass are all common. Wild celery grows well in sunny areas along the riverbank, even in strong current. Because of the steep gradient here, the plant zonation discussed in Chapter 3—from open water to dry land—can sometimes be seen in very narrow bands along the shore.

ANIMAL LIFE OF THE RIVER

Searching for animal life along the riverbank is not as easy as looking in a pond or stream. The water often gets deep very close to the shore, and the current is strong. Yet animal life is abundant where plants grow, and it is well worth the effort of your search. Since many of the organisms found here are described in Chapter 3, this section will concentrate on those specifically associated with the river environment.

Zooplankton, while present, are few in the river channel. Rotifers can sometimes be the most prevalent members of the zooplankton, preferring waters of mild current. Many of the same groups of protozoans and small crustaceans that live in ponds are also found here, being represented by species that are partial to flowing waters.

The water strider, *Gerris*, may flit across the surface in an area protected from the current. These striders can fly to a new home, being winged in the adult stage. Some other insects appear similar to their relatives upstream, but many have enlarged gills to cope with the lower dissolved oxygen levels in the river. *Hexagenia*, a mayfly, has flowing, plumelike gills with a large surface area to absorb oxygen. Sprawling and burrowing mayflies are common.

A scoop of mud will naturally contain worms or larvae of some sort, especially those that thrive where oxygen is not plentiful. Bloodworms (true midge larvae), *Tubifex* worms, some species of leech, and nematodes abound.

Some stretches of the muddy river bottom may be dotted with freshwater mollusks, especially where there are few or no rooted plants. Pill clams can occur in great numbers, as well as freshwater mussels and other mollusks. Each species rests in the water with a certain amount of its shell sticking above the mud. As it grows, a mollusk's shell enlarges, forming annual rings. The sharp point on the back of the shell, sometimes called a *beak*, is where the young mollusk began. Each shell has two halves or *valves* that are lined with mother of pearl. A large foot and two siphons are visible when the valves are open. When the mollusk wants to move, it sticks out its foot, grips the mud, and draws the shell along the bottom toward the foot.

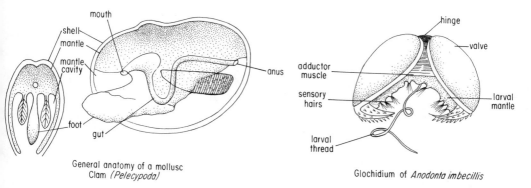

mouth
shell
mantle
mantle cavity
anus
foot
gut

General anatomy of a mollusc
Clam *(Pelecypoda)*

hinge
valve
adductor muscle
sensory hairs
larval mantle
larval thread

Glochidium of *Anodonta imbecillis*

FIGURE 5–28: General anatomy of a mollusk. Clam (Pelecypoda). Glochidium of *Anodonta imbecillis*. *Size:* .002–.02 inch (.05–.5 millimeter). (Glochidium redrawn with permission from Robert W. Pennak, *Freshwater Invertebrates of the United States*. New York: John Wiley and Sons, Inc., 1978, p. 743.)

Zooplankton, phytoplankton, and detritus are sucked into the mouth, filtered by the gills, and the waste is then eliminated from the anus. The mantle can sense touch and detect shadows, thus warning the mollusk to snap shut if danger is near.

Most surprising of all is the reproductive cycle of freshwater mussels. The minute larva, called a *glochidium* (.002–.02 inch or .05–.5 millimeter) must attach to either the gills or scales of a specific species of fish within a few days after emerging or it will perish. Once it hooks onto the host, the fish's tissues grow over and enclose it. In 10–30 days the young mussel falls to the bottom. Mussels burrow into the mud during the winter.

The river is no exception to other freshwater habitats in being replete with the streamlined shapes of fish darting through the weeds and leaping for an insect meal at the surface. Some pond species such as bass, yellow perch, and catfish are to be found. Carp are often an ecological disturbance to other river fish as they stir up the bottom sediments looking for plants, fish eggs, and small invertebrates to eat. The common sucker is another detrital bottom feeder. A fleeting shadow among the weeds may be the northern pike, voracious hunter of small animals. Weeds are also its spawning grounds.

Numerous other animals frequent the river. The mudpuppy is a powerful salamander with bushy external gills. It can grow to be 12 inches (30.5 centimeters) long, and it eats fish, insects, and other invertebrates. Painted and red-eared turtles bask in the sloughs. Snappers are probably not far off. A spotted sandpiper dips up and down along the bank. No animal can match the deft bank climbing of the mink as it looks for food. And no other is as loved as the mink's cousin, the river otter. A family of otters will often float down a river, playing, sliding down the bank, and casually hunting. Few of its prey—among them crayfish, fish, frogs, and even small muskrats—can outmaneuver the otter when it is in hot pursuit. The otter has some aquatic adaptations that are remarkably like a beaver's: flaps that close to protect the nose and ears when submerged; an oil-coated coat that repels water; and webbing on its toes. Although they can grow to be over 3.3 feet (1 meter) long, the nocturnal otters are not fond of people and are best sought out from dusk to dawn. Once you have seen a frolicking otter family with pups, you will have witnessed one of nature's greatest entertainments.

FIGURE 5–29: Mudpuppy (*Necturus maculosus*). The folds of skin covering the neck region are external gills. *Size:* to 12 inches (30.5 centimeters).

FIGURE 5–30: The playful antics and agility of the river otter help to make it a highly successful predator. *Size* (including tail): to 3.3 feet (1 meter). (*Photo by Cecil B. Hoisington.*)

Having borne witness to the antics of otters, beds of mussels, countless basking turtles, and waving beds of wild celery, the river must eventually arrive at its resting place in the sea. Sometimes well inland a river encounters the *salt wedge*, where fresh water meets salt water. Because it is less dense, the fresh water initially rides over the salt. The "wedge" of salt water beneath moves farther upstream during high tide, then recedes when the tide goes out. The mixing of fresh and salt water forms a brackish river. This changing salinity and the constantly fluctuating salt wedge create difficult living conditions. This is the *estuary*, a place of low diversity where fewer plants and animals can live, but those that can tolerate this environment are usually abundant.

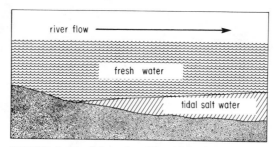

FIGURE 5–31: Salt wedge. The tidal salt water, which pushes upstream from the ocean, flows under the fresh water, which is less dense.

Mangrove swamps form in the estuaries in southern regions, and salt marshes are found in temperate climates. Planktonic blooms can be dramatic in the estuarine river as nutrients precipitate out from chemical reactions along the salt wedge. Some mussels and clams can live in the changing salinity of the bottom. And downstream appear mummichogs or killifish, razor clams, flounder, and sticklebacks. These are true denizens of the salt and brackish waters, for your journey downriver has reached the domain of the ocean.

As you ponder the swelling and breaking of waves before you, the awesome expanse and power of the sea, think back to the cool beds of fountain moss and seeping springs at the river's source. From the most humble beginnings, water meets water in seeking the lowest point, and the river grows strong to the sea's edge. There is order here. The plants and

FIGURE 5–32: Breaking waves and the broad expanse of the ocean mark an end to the river's journey. (*Photo by Michael J. Caduto.*)

animals found along the way are a response to the environmental changes that follow the confluence of many streams. The gradual birth of a river marks a journey from its upland wellspring to its destined rendezvous with the sea.

❧EXPLORATIONS AND ACTIVITIES❧

BASIC SAMPLING

Besides looking under rocks in the streambed, one of the most effective devices for catching stream insects is a screen. A screen that works well has dimensions of about 2 feet (.6 meter) wide by 2 feet high, with a long, sturdy handle attached at right angles to the long edge. In a pinch, an old window screen makes an excellent substitute. Hold the screen downstream from the area to be sampled and perpendicular to the current. Then swish the gravel and rocks upstream and wait a few seconds for your quarry to wash into the screen. Lift the screen gently and remove the animals, using either your hands or a pair of tweezers. Careful! Do not crush their delicate gills and other structures.

If you want to keep your stream insects for any length of time, place them in stream water (chlorinated tap water will kill them) and lower the container into ice water. The cold water can hold more dissolved oxygen, enhancing the survival rate of your creatures. Even so, they should not be kept longer than an hour, and always in the shade.

When your curiosity is satisfied, return the insects to a still pool at the edge of fast water, as close to where you found them as possible. This way they can seek their own optimal environment. If you place them directly into fast water they may be dashed on the rocks by the swift current.

Many of the same techniques can be used to take samples along the river's edge as were described for the pond. You may, however, want to attach a metal strainer onto a long handle to reach into the deep water along the steep banks. A plankton net can be held in the water from the bank or dragged behind a boat.

STUDYING SUCCESSION IN A LEAF BAG COLONY

You can do your own research on what kind of stream organisms prefer different substrates. Take onion bags, the kind that look like large-meshed nets, and fill them with leaves. Use different kinds of leaves in some bags, but fill a number of them with the same kind of leaves. Aspen and alder leaves work well. Fasten each bag securely to a large rock or brick to make sure it is not washed away. You can even experiment with different kinds of weights to observe colonization on different rock types, bricks, glass, and other materials. The springtime and early summer are good periods to do this because the stream benthos are active. Periodically open one of the bags to see how the colony is progressing. If you have used many bags of the same kind of leaf, you can take one colony out on each visit. Your observations will reveal a pattern of succession for insects and decomposers in the colony.

MEASURING VELOCITY

Use a tape measure to mark a length of streambed along the bank (around 22 yards [20 meters] will do) and position someone downstream at the end of the section to holler when the float finishes its course. The most convenient float to use is an orange, which floats low, roughly in the zone of maximum velocity. Time the orange to see how long it takes to travel the measured course. Repeat this at least three times and average the results.

DETERMINING DISCHARGE

Recall that discharge is the measurement of the volume of water passing a certain point in the stream over a given period of time. There are numerous ways to calculate discharge. The method used here is called the Embody Float Method, with the formula:

$$D = \frac{WzAL}{T}$$

where

D = discharge
W = average width of stream
z = average depth
A = a constant:
  .9 for sandy or muddy bottoms
  .8 for gravel or rock
L = length of stream measured
T = time for float to travel length T

FIGURE 5–33: Calculating Discharge (example)
1. Velocity = .21 meters/second (see "Measuring Velocity")
2. Velocity × A (.9 for sandy/muddy bottom or .8 for gravel/rock) = mean velocity:
   .21 m/sec × .9 (sandy stream in this case) = .19m/sec.
3. Measuring cross-sectional area: (the upper segment of this stream is shown here):

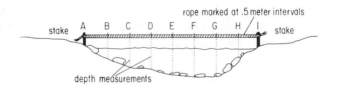

Width of stream segment (meters)		Depth measurements (meters)		Area: Width × depth (m²)
A–B	.5	B–	.26	.13
B–C	.5	C–	.48	.24
C–D	.5	D–	.50	.25
D–E	.5	E–	.59	.30
E–F	.5	F–	.62	.31
F–G	.5	G–	.60	.30
G–H	.5	H–	.21	.11
H–I	.3	I	.00	.00
				Total area = 1.64m²

4. Average cross-sectional area (csa) =
   csa (upper segment) + csa (lower segment) ÷ 2 =
       1.64m²   +   2.01m²   ÷ 2 = 1.83m²
5. Discharge = average cross-sectional area × mean velocity = 1.83m² × .19m/sec = .35m³/sec or (×60) = 21m³/min

PROCEDURE

1. Calculate the velocity using the method described in the previous exercise.

2. Multiply the velocity by the appropriate constant (A). This gives you the mean velocity.

3. Figure the cross-sectional area of the stream at the upper and lower points of the measured section using this procedure: Stretch and stake a rope from bank to bank across the stream, placing the stakes exactly at the water's edge. Measure the depth at consecutive 20-inch (.5-meter) intervals from one of the stakes. Continue until you reach the opposite bank, recording both depth and total distance from the starting stake. Calculate the area for each measured segment of the stream by multiplying the depth at that point times the width of the section (the width will be 20 inches [.5 meters] in all cases except, perhaps, the last segment when you reached the opposite bank). Add all of these areas together to get the total cross-sectional area of the stream at that point. Average the cross-sectional area at both the upper and lower ends of the measured sections of stream.

4. Multiply the average cross-sectional area times the mean velocity to get the discharge.

❧ADDITIONAL READING❧

Andrews, William A. *Freshwater Ecology*. Englewood Cliffs, N.J.: Prentice-Hall, Inc., 1972.

Barnes, James R. and G. Wayne Minshall, ed. *Stream Ecology: Application and Testing of General Ecological Theory*. New York: Plenum Press, 1983.

Coker, Robert E. *Streams, Lakes, Ponds*. New York: Harper and Row, Publishers, 1968.

Credland, Peter, and Gillian Standring. *The Living Waters: Life in Lakes, Rivers and Seas*. New York: Doubleday and Co., Inc., 1976.

Goldman, Charles R., and Alexander J. Horne. *Limnology*. New York: McGraw Hill Book Co., 1983.

Hynes, H. B. N. *The Ecology of Running Waters*. Liverpool: Liverpool University Press, 1970.

Klots, Elsie B. *The New Fieldbook of Freshwater Life*. New York: G. P. Putnam's Sons, 1966.

Pennak, Robert W. *Freshwater Invertebrates of the United States*. New York: John Wiley and Sons, Inc., 1978.

Reid, George K., and Herbert S. Zim. *Pond Life*. New York: Golden Press, 1967.

Usinger, Robert L. *The Life of Rivers and Streams*. New York: McGraw-Hill Book Co., 1967.

❧NOTES❧

1. Robert W. Pennak, *Freshwater Invertebrates of the United States* (New York: John Wiley and Sons, Inc., 1978) 12.

PART FOUR

A LOOK AT WETLAND ENVIRONMENTS

CHAPTER SIX

BETWEEN THE LAND AND DEEP WATER
Wetlands

"L et's go carefully and quietly," I whispered. "Dip your paddles in edge first so they don't splash—the noise will scare the birds and animals away before we get close enough to see them."

Everyone nodded their silent understanding and we paddled down a snaking backwater of the Mississquoi River in northwestern Vermont. "Mississquoi" is an old Abenaki Indian word meaning "at the flint"; the river was so named for the stone found nearby, which was excellent for fashioning projectile points. The cool morning air hung low over the water. Each canoe parted this gray blanket as it passed, leaving a misty wake of swirling eddies. We somehow negotiated en masse the tightly curving meanders of this little tributary, suffering only an occasional resonating bang as a misplaced paddle struck an aluminum gunnel. The narrow waters finally opened into an oval bay about twenty feet across. The canoes funneled in.

"There's no place to land the canoes along these steep banks," someone quietly protested.

"We'll have to tie them off in those alders," I said. "Be sure to have someone steady the boat as everyone else climbs out."

"What's the swelling on this plant? It looks like some sort of cancer."

"That's groundnut, a favorite food of the Indians. It grows at the water's edge, and the roots are often exposed by erosion along the riverbed," I replied.

In a moment we clambered up the slippery banks and stood atop the berm of a long dike, gazing out over an expansive, primitive-looking marsh. The dike had been constructed to raise the water table to create marsh habitat for waterfowl and other fauna. An enormous bird with a gangly neck rose silently in the distance, appearing much

FIGURE 6–1: Groundnut (*Apios americana*). *Size:* flowers .5 inch (1.3 centimeters) long, tubers .5–2 inches (1.3–5 centimeters).

like a miniature pterodactyl, one of the first flying reptiles from the later Mesozoic era. With each graceful wing beat its neck folded inward until it was neatly tucked over its back.

The coming of dawn was foretold by this brief appearance of one of its majestic messengers. The sun's first warming rays found us spellbound by the sight of the great blue heron, and slowly the haze was lifted from the marsh. "Look at meeee! . . . Look at meeee!" a red-winged blackbird boasted as he swayed on a whip-thin maple at the water's edge.*

In the predawn hours at the beginning of this excursion we had slipped our canoes off a grassy shore, each with a vision of the small patch of wilderness that lay ahead along the river and in the surrounding wetlands. There are still people who think of wetlands as mosquito-filled sloughs with tangled shrubs and twisted maples growing in pools of water where snakes and snapping turtles lurk in the oozing muck, and who see wetlands as barriers to construction and development. In fact, wetlands are invaluable natural areas, both for people and for the diverse forms of plant and animal life that make their homes there.

❧ WHAT ARE WETLANDS? ❧

Whether you are canoeing through a marsh on a misty morning or picking your way through a tangle of undergrowth while searching for songbird nests in the springtime, wetlands form a unique link between our cultured human existence and the natural elements that are our heritage in the wild. In fact, many wetlands are among the last remaining wild areas near our cities: Jamaica Bay in New York City and Tinicum Marsh just south of Philadelphia are examples. Our concern in this chapter is with the nature of freshwater wetlands, the ecological values they possess, and the effects of human actions on these precious natural resources. Over 90 percent of all wetlands occurring on the continental United States are freshwater. [1]

You may have heard the term "wetland" before. To many people it suggests a nebulous ecological concept that alludes to both water and land. This confusion is partly caused by the number of definitions that are used by people of different disciplines: a town planner will give you one definition and a wildlife biologist another, while the various levels of government each have their own definition of wetlands. Put simply, wetlands form the transi-

*Excerpted from "Wetlands, Our Great Providers" by Michael J. Caduto from Audubon Society of Rhode Island Report, September 1978. Used by permission.

tion between upland or "dry" habitats and the deep water of rivers, lakes, or the ocean. The United States Fish and Wildlife Service defines wetlands as "lands where saturation with water is the dominant factor determining the nature of soil development and the types of plant and animal communities living in the soil and on its surface."[2] The water table is usually described as being slightly below, at, or above the ground level at any given time of year. Wetlands can be flooded permanently, seasonally, or only intermittently, while some, such as bogs, are rarely flooded, but have soils that are saturated to the surface most of the time. Because they are transitional ecological zones, wetland boundaries can be difficult to locate precisely. Take a simple walk from a hillside forest down into a swampy lowland and you will see how difficult it can be to pinpoint where the upland forest ends and the wetland begins.

FIGURE 6–2: Water dominates the wetland environment, determining the plants and animals found here and the nature of soil development. The mound of vegetation to the right is a muskrat lodge. (*Photo by Michael J. Caduto.*)

When water fills soil pores, it displaces the air that is needed by soil microbes; this condition inhibits the decaying action of bacteria and fungi. As a result, organic matter accumulates, mostly in the form of dead plant remains, giving soils their peatlike or mucky character. Only plants that are adapted to the low oxygen levels of these saturated soils can grow here. Some plants, like water lilies, are found only in wetlands, while others, such as red maple, which may also grow in the uplands, have a special tolerance for wet soils.

The lower limit of wetlands is generally accepted to be where the water reaches a depth of about 6.6 feet (2 meters). This is the deepest water in which emergent plants can grow. Below this point, water, and not air, is the primary habitat for plants. Many animals, too, share the two-meter limit. Beyond this depth, whether in lakes or rivers, are found *deep-water habitats.*

At this point, a question has possibly come to mind; "If this two-meter limit is the same depth given in Chapter 3 for the limit of plant growth in a pond, and if ponds usually have plants growing all the way across the bottom, then aren't most ponds also wetlands?" An excellent question! The answer is yes. Up to this point we have looked at many freshwater environments where the water is shallower than two meters. All of these environments are considered wetlands. Included are ponds, the shallow waters along lakeshores (here the two-meter mark often corresponds with the edge of the littoral zone), most streams, and the shallow edges and backwaters of rivers.

FIGURE 6–3: This backwater of a large river is a productive, marshlike environment that helps to support life in the river channel. (*Photo by Michael J. Caduto.*)

❧THE ORIGIN OF INLAND WETLANDS❧

On the plains of the Dakotas a raft of migrating waterfowl rests in one of countless prairie potholes. A raccoon forages for food to store fat for the long winter months along the edge of a beaver marsh in the Adirondack Mountains of New York. A turtle soaks up the waning warmth of fall while basking on a cypress knee beneath graceful strands of Spanish moss in a South Carolina swamp.

Why have such diverse wetlands formed where they are? What has

FIGURE 6–4: Beavers and their dams are prolific producers of ponds. *(Photo by Mark Council.)*

caused some bodies of water to be deep and steep-sided, with little wetland vegetation forming along the shore, while others are shallow, with extensive swamps and marshes at the water's edge? How have some shores come to be rocky, while others consist of sand or mud?

Wetland basins have formed by the same forces that created lakes and ponds, as described at the beginning of Chapters 3 and 4. The reader is referred to those sections for definitions of these terms. Following is a summary of the major agents of wetland formation and notes on the specific kinds of basins and conditions they create:[3]

- Glaciers
    - kettle holes
    - morainal depressions and dams
- Rivers
    - oxbows
    - scrolls
    - deltas
- Landslide dams
- Beavers, alligators, and other animals
    - dams
    - ponds

- Wind
    dune swales
- Permafrost
    temporarily and permanently frozen soils (e.g., tundra)
- People
    impoundments (dams, dikes)
    excavated basins (farm ponds, gravel pits)
- Tectonic forces
- Volcanic forces
- Solution of bedrock
    sinkholes

Glaciers are responsible for most wetlands in Canada and the northern third of the United States. Both in this region and in more southern locales, many of our present wetlands exist in deep basins that have gradually filled in due to the accumulation of plant remains during ecological succession.

If you could stand next to a freshwater marsh for several hundred years, you would witness a gradual metamorphosis. From the edges inward, the open water, with its microscopic plants and animals, would gradually fill in with silt and dead plant remains. Submerged plants, such as the common coontail and naiad, would inhabit the deeper areas. In these accumulating sediments would root water lilies, marsh plants, then sedges, rushes, and grasses. Eventually, depending on the predominant plant communities in that region, a fairly stable association of plants might develop here, such as a swamp of shrubs and trees, a meadow, or a wet prairie. There is little evidence to suggest that wetlands evolve into dry habitats naturally; they typically persist as swamps or other advanced stages of wetland succession.

Human or natural agents may retard, stop, or even reverse this sequence of events over time. The rate of succession, and the kinds of plants that take root in each *seral stage* of succession, depends on local climate, geology, drainage patterns, water chemistry, water depth, and water level fluctuation.

❧FROM MARSH TO FOREST: WETLAND TYPES❧

The richness and variety of life found in wetlands is unsurpassed, and a visit to a marsh or swamp, a bog or prairie pothole, will reward you with unending new discoveries of nature's most delicate work. These wonders are revealed in a look at the different kinds of wetlands.

A good way to experience life in a wetland is to carefully and quietly walk, wade, or canoe into a spot where there is a mixture of open water and vegetation. Station yourself so that you are well camouflaged, then wait for the scene to come alive around you. You will need a hand lens, binoculars, and hip boots or waders, though some people prefer old sneakers during the warm season. And don't forget bug repellent!

As you read the following descriptions of the different kinds of wetlands, keep in mind that these are commonly used, popular names. Wetland ecologists, under the aegis of the United States Fish and Wildlife Service, have devised a standardized classification system for wetlands in the United States.[4] This classification includes a number of levels. The most general level is the *system*, of which there are three for freshwater habitats: *riverine* (rivers), *lacustrine* (lakes), and *palustrine* (marshes, meadows, swamps, bogs), as well as two for coastal habitats: the *marine* and *estuarine* systems. Palustrine originates from the Latin *palus*, meaning marsh, and the Greek word *pēlos*, referring to clay, mud, or a bog. Over 94 percent of all wetlands in the lower forty-eight states are palustrine. These systems are divided into *subsystems, classes,* and *subclasses.* Figure 6–5 compares some common wetland types and their technical names under this classification system.

Freshwater Marshes

The deepest edge of the marsh is the home of white and yellow water lilies, wild celery, chain pickerel, and the elusive lunker bass, which lurks amid the submergents. Here the endangered Florida manatee feeds on the prolific greens of the water hyacinth. In the marshes or "prairies" of the 680-square-mile (1,761-square-kilometer) Okefenokee Swamp in southern Georgia,

FIGURE 6–5: Common and technical names for freshwater wetlands.

Common Wetland Type	Technical Class Name
Open Water	Aquatic Bed
Marsh	Emergent Wetland
Wet Meadow	Emergent Wetland
Bog*	Moss-Lichen Wetland
	Emergent Wetland
	Scrub-Shrub Wetland
	Forested Wetland
Swamp*	Scrub-Shrub Wetland
	Forested Wetland

*Technical names for bogs and swamps depend upon the type of vegetation that dominates the wetland.

large floating mats of peat and vegetation, called *batteries,* can be found. Gases formed from decomposition of peat during low-water periods cause the organic materials to become buoyant and float free when the wet season comes. Coots and diving ducks frequent the deep marsh.

It is marshes such as those of the prairie pothole region that are most valuable for waterfowl as feeding and nesting areas during the spring and summer, and as havens to find food and a place to rest during migration. In the shallow marshes the female red-winged blackbird, looking like a large sparrow, calls from cattail stalks and then swoops down to tend her nest of three or four reddish-brown-streaked azure eggs. Red-wings are fond of wet places with patches of dense shrubs, cattails, or other emergent plants. Males can be seen from March on in the springtime, courting a mate from a conspicuous perch. Females arrive from the southern wintering grounds soon after the males come north to stake out their territories in the marsh. When performing the mating call, males assume a hunched-over position, open their wings and tail slightly, and expose their brilliant red shoulder patches or epaulets. Then they belt out a loud "Konk-la-ree" that to me sounds aptly like "Mate with me!" Sturdy nests are constructed of cattail

FIGURE 6–6: A red-winged blackbird nest amid the cattails.
Size: eggs measure .7 by 1 inch (1.8 by 2.5 centimeters).
(*Photo by Michael J. Caduto.*)

leaves, sedges and other reeds, rootlets, and grass, and they are often found low over the water in thick greenery. A red-wing's diet consists mostly of weed seeds and a smaller portion of insects. Once the shorter, colder days arrive in the fall, red-wings congregate in large masses and then migrate to their wintering grounds in the southern states.

Muskrats use the cattails as food and to build their lodges of up to 4 feet (1.2 meters) high and 6 feet (1.8 meters) across. These large rodents can reach three pounds in weight. They have an 8-inch (20-centimeter) tail that is used as a rudder as they swim, with the aid of the partial webbing on their hind feet. Where there is one muskrat, there are usually many; they can produce up to three litters every year! Muskrats eat other plants besides cattails, such as water lilies, pondweeds and arrowheads, and they are especially fond of mussels. The nutria, a rodent that was introduced from South America into southern marshes of the United States in 1899, is now common throughout the South. These large rodents can become so numerous that massive "eat outs" result where vegetation is stripped from sections of a marsh, sometimes harming populations of fish, amphibians, and other life that relies on these plants for cover.

With the abundant life that is found in fresh marshes, from the microscopic plankton to the northern harrier (or marsh hawk) that waits for a meadow jumping mouse to stray along the shore, marshes are among our most productive inland wetlands. To those people who frequent the marsh, the harrier's silent, gliding flight is a common and welcome sight. It will often appear suddenly as it follows the contours of each dip and rise of the land and plant cover. If the harrier's flight pattern is not familiar to you, then its conspicuous white rump patch, slender body, long tail, and blackish wing tips are good keys. Males are very pale compared to the darker females. Sometimes it will hover briefly and then swoop down to prey on its chief food of mice, rats, and other rodents. Other mammals, birds, frogs, and insects are also eaten, along with marsh rabbits in the southern marshes

FIGURE 6–7: Northern harrier (Marsh hawk). *Size:* body, 17.5–24 inches (44.5–61 centimeters); wingspread, 42 inches (106.7 centimeters).

where the harrier overwinters. The lengthening days of March find the male harriers arriving in their mating grounds before the females return. Males will court the females with a dramatic display of swift swoops down close to the ground, then rising and diving again many times in succession. Females lay 4–6 whitish eggs in a nest of sticks, grasses, and seeds built close to the earth or low over the water. Unlike other hawks, northern harriers commonly land on the ground.

Plant cover in the marsh can vary from swaying beds of reeds and rushes to dense floating stands of duckweed and watermeal. The soil is composed of sand, silt, or the soft, black remains of dead plants and animals. Even during the winter, when the painted turtles, mudpuppies, and bullfrogs have all burrowed into the mud to hibernate, a frozen cattail marsh provides cover for cottontail and marsh rabbits, ring-necked pheasants, and wild turkeys.

As noted earlier, marshes and ponds are different names for similar environments. The reader is referred to Chapter 3 for a detailed description of the plants, animals, and living conditions in the pond and marsh.

WET MEADOWS

Grasses, rushes, and sedges dominate the seasonally flooded land of the wet meadow. Meadows are favorite spring haunts of robins, raccoons, and song sparrows. The nasal "peent" of male woodcocks and the eerie, winnowing call of snipe can be heard in late March to April. In late fall, winter, and early spring, shallow standing water frequently covers the meadow, and

FIGURE 6–8: Grasses, sedges, and rushes abound in wet meadows. (*Photo by Michael J. Caduto.*)

dabbling ducks such as the mallard, shoveler, and pintail use these areas for resting and feeding. As the plants resume growth in the springtime, the even, new growth on clumps of tussock sedge appears like green crew cuts. In some temporarily flooded meadows the water table drops below the surface early in the growing season, but the soil remains saturated. If a meadow is intermittently flooded, there may be standing water present for only brief periods of time, and not at all in some years. Meadows are commonly found along the floodplains of streams and rivers and in low, mucky black patches of grazing land.

You can often find a diverse plant community growing in wet meadows. Small bedstraw and tearthumb may pull at your legs as you enter. Your hand lens will reveal the rows of decurved barbs that hook shoelaces and pants legs. Smartweed and three-way sedge may be growing nearby. A little farther along you spot marsh Saint Johnswort and swamp candles. The spring blooms of cowslips or marsh marigolds are a larder for early insects.

SWAMPS

"Who cooks for you—who cooks for you all," cries a barred owl in search of a mate among the crowns of dormant red maples. The most common owl in many wooded areas, barred owls are year-round denizens of the swamp. Unlike the glaring yellow, saucerlike eyes of the great horned owl, barred owls have dark eyes and a rounded head that give these birds a gentle demeanor. They are, however, capable of producing blood-curdling screams

FIGURE 6–9: Barred owl. A common and welcomed sight in the wooded swamps, barred owls will often afford a close look to a cautious observer. *Size:* body, 17–24 inches (43.1–61 centimeters); wingspread, 44 inches (111.8 centimeters).

that pierce the night air. Barred owls are often seen during the day or at dusk. They frequently make short flights from a roost around 40 feet (12.2 meters) up in a tree to another branch nearby, coming in low and swooping up as they land on the new perch. If an observer moves slowly and quietly, these owls will often allow an approach to within 30 feet (9.1 meters) or so. A distinctive trait of the barred owl is its set of lateral bars on the upper chest or neck region and vertical lines on its breast. Although they have been recorded eating screech owls and even long-eared owls, the barred owl subsists mainly on mice, occasionally eating other mammals, birds, amphibians, insects, reptiles, and even fish. Barred owls usually nest in hollow trees or old red-shouldered hawk nests, often returning to the same nest for many years to lay usually two pure-white eggs. Some pairs of barred owls and their descendants have nested in the same patch of woods for over thirty years! Courtship occurs in March, and the offspring of these owls will be born in the wetland forest during the lengthening days of late winter and early spring, when the skunk cabbage heads are poking up through soggy snow.

Although the temperature may hover near freezing, tiny flies and an occasional honeybee will rest within the skunk cabbage spathe, a hoodlike protective covering for the flower, to store heat between their foraging flights. The skunk cabbage, relative of the tropical *Philodendron* and *Dieffenbachia*, can maintain a temperature of 71.6°F (22°C) as long as the air remains above freezing. As the skunk cabbage matures, the swamp will be alive with another seasonal denizen, the lovers of avian life who aim their double-barreled binoculars into the treetops to glimpse the fleeting feathers in this, the habitat that harbors the most diverse populations of spring songbirds. Wooded swamps are the breeding grounds of thrushes, vireos, numerous flycatchers, and eye-wearying warblers.

Wooded swamps, also known as *forested wetlands*, are areas where the soil is saturated and often inundated, and trees form the dominant cover. This is the most abundant type of wetland, comprising over one half of the total wetland acreage in the continental United States. The soils of wooded swamps are rich in organic matter: In some northern swamps peat deposits may be as thick as 20 feet (6 meters) or more. Shrubs typically form a second layer beneath the forest canopy.

On a walk through the southern swamplands you may see the bright ruddy crest of the nearly crow-sized pileated woodpecker as it searches for insects beneath the bark of a dead water oak. These striking woodpeckers are secretive residents of the wooded bottomlands. When looking for the pileated woodpecker, an occasional loud "ka-ka-ka" may be heard, but the unmistakable signs of these birds are the large, oval-shaped excavations that they produce in dead trees while searching for their favorite food—ants and other insects. These holes can range up to 1 foot (30 centimeters) or more in height, in line with the tree trunk, and 4 inches (10 centimeters) across.

FIGURE 6–10: The pileated woodpecker searches for insect food in the treetops. *Size:* body, 16–19.5 inches (40.6–49.5 centimeters). (*Photo by Cecil B. Hoisington.*)

FIGURE 6–11: The true function of cypress "knees" is still a mystery; they are thought to aid in respiration. (*Photo by Cecil B. Hoisington.*)

They are more frequently chiseled during the late winter to early spring, when insect food that can be obtained by other means is scarce. The power used during these arboreal excavations is evident in the piles of wood chips dropped at the base of trees, with some larger chips measuring 2–3 inches (5–7.6 centimeters) across! Squirrels, mice, and tree-nesting birds will often inhabit old pileated woodpecker holes. The male pileated uses a drumming call during the mating season of April and May. Nest holes, in which usually four eggs are laid, are carved around 45 feet (13.7 meters) up into dead trees, mostly in lowland hardwood forests and swamps, or in woodlots with a stream close by. The opening usually faces south or east. Although pileated woodpeckers require large tracts of forest and are usually reclusive and hard to sight when they are being sought out, they will sometimes appear suddenly and close by. I saw one bird as it casually chipped away at a pine tree on a busy street in a town of several thousand people! And these spectacular birds frequently delight guests at the Okefenokee Swamp in Georgia when they appear during the winter at a feeding station near the visitor's center.

Some common trees in the southern swamp are the loblolly bay, the tupelo gum, and the bald cypress, with its "knees" protruding from the reflective waters. These knees are thought to serve a respiratory function. Common in the northern Atlantic states are red maple; black gum, which is the first tree to turn red in the fall; larch; the Atlantic white cedar; black ash; pin oak; and swamp white oak. Willow, blueberry, and speckled alder are shrubs that often grow in the understory in this region.

In the coastal regions of the Carolinas are found upland evergreen shrub swamps and forested wetlands called *pocosins*. Bay and pond pine are important trees, with leatherleaf, wax myrtle, fetterbush, and titi growing in the understory. The national classification system refers to shrub swamps as *scrub-shrub wetlands*. Some common shrub swamp plants in the Northeast include red osier dogwood and silky dogwood, swamp rose, pussywillow, winterberry, spicebush, sweet pepperbush, elderberry, and swamp azalea. Beneath the shrubs frequently is a rich community of ferns and herbs: jewelweed, royal fern, cinnamon fern, and sensitive fern are just a few.

The white-footed mouse and red-backed vole can be found scurrying among the sensitive fern and *Sphagnum* at the base of swamp maples. A red fox may happen by and send these small creatures into their burrows. Abundant food and cover attracts deer, ruffed grouse, and tree-cavity nesting birds like the common and hooded mergansers and the wood duck—the male woody with its resplendent plumage. Mergansers are diving birds that feed primarily on fish in lakes and rivers near the swamplands where they often nest. Wood ducks and hooded mergansers are frequently seen together during the breeding season. Woodies produce a nest of 10–15 eggs in the hollow of a tree from 3.3 feet (1 meter) to 50 feet (15.2 meters) off the ground. Their beautiful, ivory-colored, somewhat glossy eggs yield young ducklings with sharp claws that help them to climb out of the nest hole. Their first step into the real world sends them plummeting from the nest to the ground, where they are usually unharmed by the experience. A wood duck duckling has next to make its way to the relative safety of a pond, marsh, or lakeshore. But since the nest tree can often be thousands of feet from the shoreline, the ducklings are easy prey for foxes, mink, and other predators. Once in the water, they are vulnerable to the jaws of snapping turtles and large, raptorial fish like pickerel and pike. At one time even the adults may have fallen prey to the duck hawk or peregrine falcon. But this, our swiftest of falcons, is now few in number due to nesting failures caused largely by the effects of the pesticide DDT, although efforts to reintroduce peregrines into their former ranges are beginning to succeed. A wood duck's food is taken mostly at the water's surface and on land: insects such as beetles and mayflies, fish, tadpoles, seeds, leaves, berries, and—among its favorite foods—beechnuts and acorns. A year-round resident of states in the South, wood ducks are seasonal nesters in the North. In the early 1900s their numbers were decimated by hunting, the draining of swamps to create croplands, and the destruction of nesting trees by logging operations. Thankfully, wood ducks are returning to many formerly inhabited bot-tomlands due to strict hunting regulations, the return of forests to many abandoned agricultural lands, and the successful use of artificial nesting boxes.

FIGURE 6–12: ♂ Wood duck. Due to wise management practices, which followed decades of dramatic population decline, the beautiful wood duck is once again becoming a common sight in some regions. *Size:* body, 17–20.5 inches (43.1–52.1 centimeters).

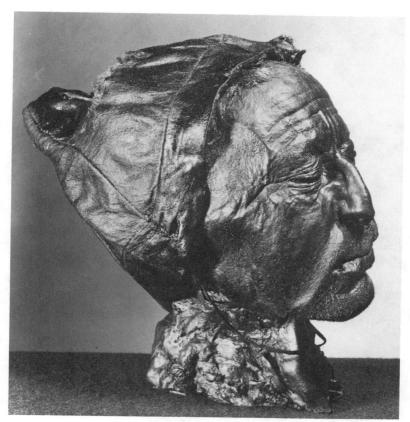

FIGURE 6–13: Tollund man. (*Photograph courtesy P. V. Glob, Director, Danish National Museum, Denmark.*)

BOGS

Two Danish men were harvesting bog peat one day in May of 1950, when one called out to the other. The two of them stood staring at the body of a man with a noose around his neck, surrounded by the peat. The police were notified; they suspected that something was awry and so consulted an archaeologist. Finally the body, now known as Tollund man, was dated back to the Iron Age, 2000 years ago.[5] As old as he was, Tollund man was perfectly preserved; his beard, for instance, clearly showed several days' growth.

For most of us, the bog may be our only chance to attain corporeal immortality. Forming in glacial kettle holes or in other places where surface drainage is congested, bogs develop best where soil and water temperatures are low. These conditions, combined with the dearth of oxygen and mineral

FIGURE 6–14: (*above*) Pitcher plant (*Sarracenia purpurea*). Pitcher plants supplement their nutrient-poor diet with insects. *Size:* 8–24 inches (20.3–61 centimeters). (*Photo by Michael J. Caduto.*)

FIGURE 6–15: (*right*). Thread-leaved sundew (*Drosera filiformis*). An insectivorous bog plant. *Size:* 4–12 inches (10.2–30.5 centimeters).

nutrients, result in incomplete decomposition of organic matter—dead plants build up to form a fibered, spongy soil, unlike the black muck of the marshes and swamps. This *peat* can fill basins to a depth of over 40 feet (12.1 meters) deep. Bogs, depending upon the dominant plant community, are sometimes classified as *moss-lichen wetlands*, scrub-shrub wetlands, or forested wetlands.

Bogs are unique and fragile environments harboring many plants that are restricted only to bogs, especially in the South. A bog often begins forming at the edge of a body of water, with *Sphagnum* moss and sedges creating a floating mat. *Sphagnum* contributes to highly acid growing conditions. Sedges, cotton grass, and some lichens can tolerate the wet, acid conditions of the bog mat, along with the white-fringed orchis. Bogs are the home of many beautiful and delicate orchids: the white-colored lady's tresses, rose *Pogonia*, yellow lady's slipper, and both *Calopogon* and *Arethusa*, with their flowers ranging from pink to magenta.

And here are found some insectivorous plants that supplement their nutrient-poor diets with insects: horned bladderwort, pitcher plants, and sundew are widespread, while the Venus' flytrap can be found in the southeastern United States. Not all of these insect-eating plants are as well-known or dramatic to watch as the Venus' flytrap, which quickly snaps shut and traps unwary insects. The bladderwort has ingenious insect-trapping sacs on its underwater leaves that are deflated when at rest. When a small animal, such as a nematode, swims near a bladder and touches one of the trigger hairs near its opening, the sac suddenly inflates and the hapless creature is sucked inside by the rushing water and then held tight by a trapdoor. The insect is then digested. Once its meal is finished, the sac deflates and the trapdoor opens in preparation for another catch. The pitcher plant has red veins and a scent that guides insects into the opening. The hairs on the inner lip of the hood point downward, making it difficult for insects to travel out of the "pitcher." Once they are trapped in the fluid of the pitcher, they are slowly digested by enzymes. Nevertheless, the female of one species of mosquito (*Wyeomyia smithii*) lays eggs in the pitcher plant and the larvae develop there, curiously immune to the digestive juices. Sundews possess drops of clear, sticky fluid on the ends of hairs in which small insects founder. The sundew's Latin name, *Drosera*, comes from the Greek *droseros*, meaning dewy. Slowly the hairs, and the sticky pads to which they are attached, fold in toward the center and entrap the insect, which becomes food. When the insect is digested, the hairs gradually unfold, ready to dine once again.

Buckbean and water willow are two plants that commonly grow at the edge of the bog mat. Cranberry creeps over the bog, its roots binding the moss into a more cohesive substrate, along with those of leatherleaf. Other

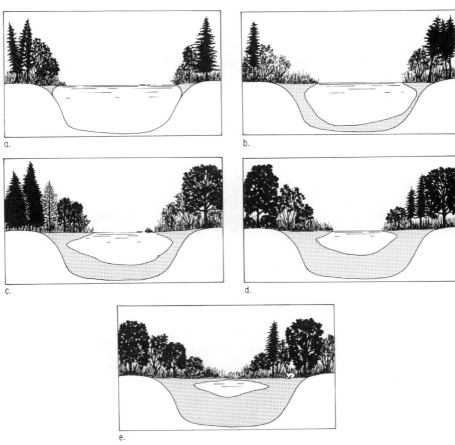

a.

b.

c.

d.

e.

FIGURE 6–16: Ecological succession in a bog. The basin is mostly open water where drainage is congested. (*a*) Some sedges and *Sphagnum* moss are beginning to grow out from the edges. Peat continues to accumulate and the area of open water begins to shrink as the bog mat develops further. (*b*) and (*c*) Older sections of the bog are supporting the growth of some low shrubs. The edge of the bog mat is floating, which gives it the characteristic "quaking" habit found in young sections of the bog. As the peat builds up and becomes more consolidated with the further accumulation of plant remains, the soil becomes better aerated and is capable of supporting decomposition by aerobic fungi and bacteria. Nutrients are more available than in the young bog mat where peat is saturated and growing conditions are anaerobic or nearly so. Eventually, large shrubs and trees grow in the oldest sections of the bog (*d*), and the open water becomes closed off completely (*e*). The bog continues to develop toward a swamp.

common shrubs include sweet gale, sheep laurel, bog laurel, bog rosemary, and Labrador tea. Sweet gale improves its diet in the nutrient-poor bog soils by fixing nitrogen from the air. Labrador tea is thought to have survived in southern locations as a relic of the glacial period. Many bog shrubs—Labrador tea, bog rosemary, and bog laurel among them—have fleshy leaves with edges that are curled under, curiously similar to the adaptations found among the leaves of drought-resistant plants of arid climates. The decurved edges decrease evaporation from the *stomata*—small openings on the underside of terrestrial plant leaves through which gas exchange and water loss occur. Why are these adaptations found among shrubs growing in saturated peat? The conditions of low oxygen and high acidity inhibit the normal physiological functioning of roots which allows them to absorb water for use by the plant. For these shrubs, the bog creates an artificial drought amid water aplenty.

Eventually, as the peat builds up and consolidates, taller shrubs and trees take root in the bog mat: black spruce and larch in the Canadian muskegs and Atlantic white cedar in coastal bogs. Deciduous trees such as red maple do not grow well in the bog until the peat is firm and has begun to decompose.

Because of the scarcity of valuable food plants, few animals frequent the bog. Water shrews and masked shrews are common, along with white-tailed deer, otters, and minks. Shrews have poor eyesight, but they orient well to find their prey with a keen sense of touch, for they are voracious feeders. Masked shrews have pointed noses and a thick pile of short hair that is blackish in winter and lighter in summer, with a gray to light brown coat underneath. They are around 3.9 inches (10 centimeters) long, including their tail, which is around three fourths of their body length. While tunneling through the moss, masked shrews, which are especially active at dusk, can eat up to three times their own weight in one day! As long as they are small enough, most animals and even some plants are fair game: spiders, slugs, snails, beetles, worms, seeds, and even other masked shrews.

At around 4.3 inches (11 centimeters) in length, water shrews are slightly larger than masked shrews and, as their name implies, they are good swimmers. Their hind feet have partial webbing between the third and fourth toes and stiff hairs that help to propel them through the water. They can swim underwater, "walk" along the bottom while submerged, and have even been seen skittering over the water along the surface film for a distance of close to 6.6 feet (two meters)! Being primarily nocturnal, water shrews eat fish eggs, aquatic insects such as mayfly nymphs, and caddisflies; they are even swift enough swimmers to catch small fish. Their fur is blackish during the winter, brownish during the summer, and is tipped with white, which gives their coat a frosted appearance.

Another small mammal of the bog, the southern bog lemming, has a coarse coat of long, brownish-black hair on its back which appears silvery on the sides and belly. This is one of the few rodents that is found especially in *Sphagnum* bogs.

The avian community is not as diverse when compared to other wetland types: black and ring-necked ducks; the northern waterthrush; golden-crowned kinglets; and song, swamp, and white-throated sparrows are common. You might see cedar waxwings or several species of warbler, such as the palm warbler, common yellowthroat, yellow-rumped warbler, and the parula or magnolia warblers.

The bog turtle, a threatened species in the United States, is an uncommon resident of the bog. This small (3.9-inch or 10-centimeter) turtle is mostly brown, with a distinctive splotch of orange on each side of its head. A

FIGURE 6–17: Bog turtle, an uncommon sight amid the *Sphagnum* moss. *Size:* 3.9 inches (9.9 centimeters).

shy animal, it will often burrow into the mud when disturbed. It eats mostly insects, such as caterpillars and beetles, and some seeds and berries. Bog turtles emerge from hibernation in April and are active through mid-fall. Males have long tails, strong front claws, and a concave *plastron*, or lower shell, which enables it to fit over the female to mate in May to June. Females have a small notch at the posterior of the plastron. Inhabitants of swamps and poorly drained marshes and meadows, as well as bogs, the females bury their clutch of 3–5 eggs in the ground. Unfortunately, these rare animals continue to dwindle in number.

Bogs are indeed a unique freshwater environment harboring many rare and unusual plants and animals. However, humankind has found bogs to be useful for harvesting peat moss, a major industry and export in Canada, and for taking peat to be dried for use as fuel. Peat harvested from Scandinavian bogs is a major source of heating fuel for homes and industry. Since decomposition is so poor in bogs, the trees that are found in the peat are usually sound and are a valuable resource for use as poles and lumber, even though they may be hundreds or even thousands of years old! The struggle between preserving bogs for their natural beauty and as homes for many unique and intriguing plants and animals, versus making commercial use of their natural products, is a microcosm of the widespread conflict between the use and development of our wetlands and the conservation of these valuable and irreplaceable environments.

❧THE VALUES OF WETLANDS❧

If you have read this far, it is obvious that you are interested in the biotic and abiotic elements of freshwater environments; you are aware that they hold certain values for plants, animals, and people. Nevertheless, we all seem to want an answer to the perennial question, "What good is it?" Many researchers have devoted their lives to answering this question for wetlands, and this section summarizes those findings.

Historically, wetlands were regarded as wastelands, places to be dredged or filled to suit our needs. As a result, the statistics regarding human impact on wetlands are staggering. When the United States was first settled by Europeans, 215 million acres (87 million hectares) of wetlands existed here. We now have only 99 million acres (40 million hectares) of wetlands left in the lower forty-eight United States. This is only 46 percent of our original wetland acreage, and it covers just 5 percent of the land surface of these states.[6] The rest have been filled, drained, or in some way altered so that they no longer possess their natural values.

When faced with wetland destruction in the past, the major concern has been with the loss of wildlife and natural habitat. However, the surge of interest in wetland ecology that began in the 1960s and 1970s stimulated research that has documented many other values and functions of wetlands. These range from such life-sustaining and economically important functions as flood control and water supply to intangible aesthetic and recreational pleasures. A report from the United States Fish and Wildlife Service, *Wetlands of the United States: Current Trends and Recent Status*, summarizes these values under three major categories: fish and wildlife values, environmental quality values, and socioeconomic values. A close look at the values of wetlands leads to some astonishing discoveries.

Fish and Wildlife Values

An abundance of food, vegetative cover, and water accounts for the great diversity of wildlife that may be found in wetlands.[7] Perhaps the most spectacular and well-known displays are the spring and fall waterfowl migrations. During these periods, the marshes and coastal plains teem with thousands of birds as they rest and feed on their twice-annual journey. Most of the fish that are important in commerce and recreation breed and grow in the wetlands surrounding the open waters; perch, pickerel, bass, sunfish, muskellunge, salmon, bullhead, carp, and walleye are examples. Southern river swamps have been shown to produce 1300 pounds (585 kilograms) of fish per acre.[8] Many insects, reptiles, and amphibians live here—frogs, dragonflies, salamanders, and turtles. Amphibians that live in the uplands, such as toads, wood frogs, spotted salamanders, and the red eft (the immature stage of the red-spotted newt) must return to wetlands to reproduce. Some wetland animals are destined to become food for green-backed herons, sandpipers, anhingas, and other birds, while plants will feed black ducks, Canada geese, and coots. Muskrats, beavers, and nutrias abide here. And many upland animals—among them raccoons, opossums, deer, and moose—frequent wetlands to feast on the bounty of plants, animals, and water. Overhead, in the treetops of a swamp, the branches are alive with the songs of warblers in the springtime. Wetlands also harbor many endangered and threatened species, like the bog turtle, Florida manatee, and American crocodile.

In the Great Plains states of western Minnesota, Montana, North and South Dakota, and up into the Canadian provinces of Saskatchewan and Alberta, prairie potholes can be seen by the thousands. Remnants of the glacial age, they number up to two hundred per square mile. So highly prolific are these wetlands that they produce one half of all the waterfowl in North America, and yet they comprise only 10 percent of the wetland area. The millions of waterfowl that nest here use the Central and Mississippi flyways during migration. Prairie potholes are irreplaceable natural resources, the producers of the prairie region, yet well over one half of our original prairie pothole acreage has been destroyed.

Environmental Quality

Water quality is frequently enhanced as water passes through wetlands. Soil microbes, plant litter, and living plants actually reduce pollution levels in water. The organic matter in wetland soils absorbs substantial amounts of nutrients and chemical contaminants. If nutrient levels become high enough, due to effluent discharged by a sewage treatment plant or agricultural fertil-

FIGURE 6–18: An anhinga dries its wings in the sunlight. *Size:* body, 34 inches (86.4 centimeters); wingspread, 47 inches (119.4 centimeters). *(Photo by Cecil B. Hoisington.)*

FIGURE 6–19: The white-tailed deer finds food and water in wetlands. *(Photo by Cecil B. Hoisington.)*

FIGURE 6–20: Prairie potholes are the most valuable inland breeding grounds for waterfowl in North America. Yet they are rapidly being drained to create new farmland, to provide flood control, for irrigation, and for other forms of development. *(Photo courtesy of United States Fish and Wildlife Service.)*

FIGURE 6–21: Floodplain wetlands enhance the water quality in nearby
waterways and possibly that of ground water. Oxygen is added to the water by wetland
plants during photosynthesis. Plant production in these wetlands supports life both in the
wetland and in surrounding open waters. (*Photo by Michael J. Caduto.*)

izer runoff, nitrogen and phosphorus may be taken up by plant roots. Bacterial microbes remove some contaminants during the processes of sulfur reduction and denitrification. The broad expanse of wetlands and the binding plant roots slow the flow of surface runoff and cause sediment to settle out. Living plants, from algae to tall emergents such as cattails, add oxygen to the water during photosynthesis. Overall, wetlands help to lower pollution levels in associated open water areas. Since surface water often seeps into the ground-water aquifers where our wells are located, wetlands may help to maintain the purity of ground water as well.

Wetlands are among the most productive ecosystems in the world. A fresh marsh is as productive as a tropical rain forest, producing around 4.4 pounds (2 kilograms) of biomass per 11 square feet (one square meter) each year! Diatoms and other algae, detritus, and macrophytes provide the base of wetland food chains.

Indirectly, wetland plants support much of the life in open water. Every year the plants of the marshes, swamps, and floodplains die and their remains are left strewn over the soil surface. Fungi, bacteria, and other soil microbes break down this material. The annual spring floodwaters carry detritus from floodplain wetlands into the main channels of rivers, where it feeds juvenile fish and aquatic insects. These organisms in turn become food for the larger, predaceous fishes, turtles, snakes, and mammals like the otter. Thus wetlands provide food for life both within the wetland and in adjacent open water.

The plants and water of wetlands also moderate local climates.

Picture yourself standing in a parking lot on a hot summer day. The heat rises in visible waves from the black, scalding pavement. Now imagine there is a river with swampy banks nearby. Walk into the shade cast by the trees and sit along the riverbank. It is cooler here because of the shelter from the sunlight and evaporation from the leaves and open water. The air is richer in oxygen from the plants, and is even cleaner than in the parking lot because green leaves can filter some pollutants from the air.

Compared to the forests and open areas of dry land, wetlands tend to moderate the local climate. Evaporation from open water and plant leaves increases moisture in the air, and the large volumes of water present act like a heat sink, decreasing temperature fluctuation in the wetland and its environs.

SOCIOECONOMIC VALUES

When runoff within a watershed becomes greater than the outflow through river channels, water levels rise in the rivers and streams, and flooding ensues. Wetlands located along the floodplain help to store floodwater and then release it slowly back into the stream or river. This prevents flooding and property damage downstream. Due to the meandering habit of gently flowing rivers, and the vegetation found growing along the banks, wetlands also help to prevent flooding and erosion by slowing the flow of water and dampening wave action. Plant roots aid further by holding the soil in place. In this way, wetlands act as buffers during heavy rains and spring thaws by reducing the maximum rate of runoff and delaying the flood crest in downstream areas. Wetlands not only help to protect the quality of our ground and surface water, they provide water supplies for irrigation and drinking as well.

The plants and animals found in wetlands also provide us with a bountiful natural harvest. Suppose you were sitting down one evening to a dinner of trout, wild rice, and cranberry bread. A crackling fire and the glowing flames on the hearth warm you, inside and out. The compelling aroma of blueberry pie wafts from the kitchen, making it hard to forget the dessert to follow.

The items in this scene that might have been harvested from wetlands include timber to build your home, fire, and furniture; the wild rice, trout, blueberries, and cranberries. Wetlands also supply us with peat as a fuel or to enrich garden soil. The fish and wildlife found in wetlands provide resources like food and furs.

Wetlands are also valuable recreational areas. Almost everyone indulges in some kind of wetland-related recreation at one time or another. Enjoyment of water sports such as boating, fishing, swimming, and skating is

FIGURE 6–22: Wetlands are natural places to go for recreation and environmental study. (*Photo courtesy of the Audubon Society of Rhode Island.*)

enhanced by their proximity to wetlands. People are attracted to wetlands because of their wild beauty, making them ideal areas for such fast-growing pastimes as bird-watching and nature photography.

Because wetlands are unique ecological environments that harbor many specialized plants and animals that cannot fulfill their needs in other habitats, these are ideal places for education and scientific research. The fascinating life in wetlands can be studied at any level, from kindergarten to graduate school. Wildlife is often very conspicuous and easily captured for study. Animals may be used to show a group of preschool children the external differences between a turtle and a frog, or to conduct graduate study on the physiological adaptations of these animals to life in the wetland.

Paleolimnology is a fascinating science that has taught us much of what we know about past climatic changes and accompanying responses among natural communities. The remains of plants and animals have been found where decomposition is poor, in the ancient strata of bogs, marshes, and lakes: diatom frustules, pollen grains, insect and other animal remains. Scientists use the process of *radiocarbon dating* to date these remains, enabling them to draw a picture of the historical progression of life on earth and in local areas. As plants and animals grow, they incorporate carbon-14 (C-14), a natural element of cosmic origins, into their tissues. Upon death, the C-14 begins to break down at the rate of 50 percent every 5570 years. This is called its *half-life*. By measuring the remaining C-14 in the organic remains of plants and animals, the date of death can be determined.

The aesthetic value or natural beauty of wetlands will become increasingly important as the neighboring farms and woodlands succumb to the encroachment of house-lot subdivisions, highways, and commercial and industrial developments. There is a dire need for the open space and wildness they provide in congested areas. Wetlands are ideal places to relax and renew a personal relationship with nature.

THE MANAGEMENT OF WETLANDS: YESTERDAY AND TODAY

Early wetland alterations were on a small scale—the creation of ponds for farm water supplies, construction of canals to convey water for use in driving water wheels, or creation of ice ponds needed before refrigeration. In recent times, alteration and destruction have been more widespread and more profound. Wetlands have been destroyed by physical actions such as draining, dredging, and filling, and by pollution with excess nutrients, toxic wastes, and sediments. Draining wetlands for agricultural development has been responsible for the loss of 87 percent of the wetlands destroyed between

FIGURE 6–23: Wetland destruction continues apace. These are
flat environments that are easily filled for development.
(*Photo by Michael J. Caduto.*)

the mid-1950s and the mid-1970s.[9] Urban development, such as construction for homes, commercial properties, and highways, has accounted for the loss of another 8 percent. The remainder has been lost to other developments, such as irrigation, flood protection, and dredging navigation channels.

Although we have created more ponds and lakes than were destroyed during this twenty-year period, the overall loss has been 9 million acres (3.65 million hectares) of wetlands, an area roughly twice the size of New Jersey. We are decimating nearly 500,000 acres (202,500 hectares) of wetlands each year. This amounts to destroying an area of wetlands as large as Delaware every three years. Since the time of settlement in the United States, some states have suffered frightening losses of their original wetland acreage. For example, the following wetland resources are gone:[10]

- 91 percent of California's original wetlands
- 99 percent of Iowa's natural marshes
- 71 percent of Michigan's original wetlands
- 50 percent of Louisiana's swamps (forested wetlands)
- 50 percent of Connecticut's coastal marshes

It is not only the major alterations that decimate wetlands; the cumulative effect of the destruction of thousands of small wetlands or pieces of wetlands is substantial.

FIGURE 6-24: Osprey. *Size:* body, 21-24.5 inches (53.3-62.2 centimeters); wingspread 4.5-6 feet (1.4-1.8 meters). (*Photo by Cecil B. Hoisington.*)

CREATIVE PROTECTION AND MANAGEMENT OF WETLANDS

The cold, damp air of the early morning nipped at my hands as I clutched note pad and paper. "Keyerrr, keyerrr," cried an osprey, its voice reverberating over the marsh, shattering the silence. Angled wings carried it into a stand of dead black gums about 45 feet (13.7 meters) tall. The osprey spread its talons and, with a resounding crack, broke off a three-foot branch for its nest.

This is not a scene at an age-old marsh of pristine natural splendor; it was witnessed during the spring of 1978 in the Great Swamp of South Kingstown, Rhode Island. Construction of an impoundment by the state Division of Fish and Wildlife transformed a 140-acre (56.7 hectares) forested swamp into a complex mosaic of open water, small islands, dead woody plants, and marsh. Not all human alteration of wetlands is destructive; creating wetlands to provide wildlife food, cover, and nesting sites has become a common management practice.

Governments and planners have begun to recognize the natural values of wetlands and are increasingly considering these in their work. Around 50 percent of the states have now passed some form of legal protection for wetlands, especially coastal estuaries. Some creative wetland management practices have been initiated.

In the early 1970s, the Army Corps of Engineers (COE) investigated the feasibility of flood-control engineering in the Charles River Watershed, near Boston, Massachusetts, to protect downstream areas. The construction of a large reservoir as well as walls and dikes were both considered. During the Charles River study, however, COE also looked at the possibility of leaving the natural flood-control value of the unaltered floodplain wetlands intact, an option that was the least expensive of the three. Observations made during a 1968 flood were the key factor in their ultimate decision to retain 8500 acres (3443 hectares) of wetlands along the Charles River in their natural state. During the flood, the water was stored and slowly released from the floodplain; it took four days to pass through this area. This prevented a rapid release of floodwater and controlled flooding downstream. On the other hand, floodwater rushed through the lower, urbanized section of the river in a matter of several hours. With the acquisition of the last of these floodplain wetlands in 1983, the COE's "Natural Valley Storage" project was completed.

This landmark undertaking to preserve the floodplain wetlands for their natural flood-control value was the first major program of its kind. It stands as a model of what can be accomplished if an objective study is made and the current knowledge of wetland values is creatively applied.

FIGURE 6–25: (*Photo by Michael J. Caduto.*)

Wetlands are crucial to plant and animal life, and they are crucial to humankind. The role of wetlands as sentinels, guarding the quality and abundance of the water we need to live, is one that cannot be taken for granted. Increasingly, the importance of wetlands in the total ecological picture becomes clear.

And what is the value to the human spirit of the intangible gifts from wetlands? When we are inspired by the feeling of timeless abandon evoked by the sight of a marsh through the early morning haze, or by the gliding wings of a marsh hawk, one thing is certain: As wild areas that have not yet been despoiled by human action, wetlands are vital to our spiritual ties with nature.

❧EXPLORATIONS AND ACTIVITIES❧

NATURALIZING

Naturalizing is an exciting way to experience a natural area from the perspective of being a part of the environment, and not apart from it. This technique can be used by naturalists, artists, writers—anyone who wants to become closer to nature in spirit and understanding—to enhance and deepen their experiences in the out-of-doors.

Choose a wetland you want to visit and find a spot where you can be comfortable for an extended period of time—next to a tree, in the soft grass, in the sun or shade. Locate yourself downwind from the area to be studied so your scent will not spook the animals. Wear inconspicuous, muted earth colors. Settle into your spot and look around you. Study the plants, rocks, water, and make a mental map—this may help you to become more observant. Now sit, quietly and patiently, for a long period of time, perhaps an hour or two. Try not to move at all, and if you must, move very slowly and smoothly so that your motion is not perceptible. You may need to build up a tolerance for sitting perfectly still by beginning with shorter periods and gradually increasing the length each time.

As you sit, the animals will begin to accept you as part of their environs, and anything can happen. I have had a turtle swim to within six inches of where I was sitting on a rock wall near a marsh. Some people have even had birds land on their shoulder or head.

WETLAND COMMUNITY SURVEY

Nothing is better than close observation and measurement, with some discernment about why things are as they appear, as a means of learning the ecological workings of an environment. Each step of this activity will direct your observations and help you to learn more about wetland ecology.

Establish a transect line, using a compass bearing and a piece of string, running from the upland to the deepwater boundary of your wetland. Choose five points along this line which fall within plant communities that represent a transition between the drier and wetter extremes of the wetland. For instance, in a marsh you might sample plants in the emergent, floating-leaved, and submergent zones. At each station, make the following survey:

1. Carefully scan the wetland to get a feel for your community. What growth forms of plants do you see (e.g., emergent, floating, shrubs). Are there any animals apparent? Where is the water table? What is the soil like? Using the descriptions of wetland types in this chapter, decide whether your wetland is a marsh, wet meadow, swamp, bog, or perhaps an interspersion of several wetland types.

2. The abiotic components of the wetland environment affect what can live there. You will later use this information to compare the living conditions at the five stations. Measure, describe, and record the following:

air temperature _____

soil temperature _____

water temperature:

    a. at the surface _____

    b. on the bottom _____

sunlight intensity (percent of the ground or water surface exposed to direct sunlight) _____

water level (check one):

    a. at the surface _____

    b. above the surface _____ depth _____

    c. below the surface _____ distance _____

3. Measure and mark a 1-square-meter plot at each station. Identify and count all the plants and animals you can. Refer to the collecting techniques described for pond animals; use a strainer and collecting jar. The field guides mentioned at the end of Chapter 3 will help. Take note of the conditions where each organism is living. Also, write down the ecological niche of each plant or animal: producer, consumer (herbivore, carnivore, omnivore, parasite), or decomposer (detritivore, scavenger, saprophyte).

4. Synthesis: Tally up the total numbers of each plant and animal found and list them from the most abundant to the least. Compare the findings for temperature, sunlight intensity, and water table level at the different stations. How might these conditions affect the plants and animals found at each station? Now create a small food web for each individual station. Then draw a giant food web using felt markers and large paper, to see how energy and nutrients might pass between all the plants and animals in your wetland. A review of the ecological concepts explained in Chapter 2 may help.

Equipment needed for the wetland community survey: hand lens, thermometer, strainer (old tea strainer will do), meter stick, identification keys to wetland plants and animals, pencils and paper for taking notes, collecting jars, felt-tipped marking pens, and large sheets of paper.

BLUE-BURNING MARSH BUBBLES

Marsh muck is wet soil. The plant and animal remains found here decay as they do on land. But because marsh soil is saturated and air is almost absent, it takes highly specialized life forms to accomplish the vital task of decomposition.

The bacteria and fungi of the marsh break down organic remains into smaller elements which are nutrients for other marsh life. These anaerobic bacteria live on the chemical energy stored in detritus. Gases such as ammonia, hydrogen sulfide, and methane are given off as this digestion occurs. Hydrogen sulfide creates the rotten-egg smell of marsh gas; methane is highly flammable. One researcher calculated that a shallow, productive marsh in a temperate region can give off 90 cubic feet (2.6 cubic meters) per acre each day of highly combustible marsh gas.

Catch these rising bubbles and you can make a marsh torch. Take a one-quart plastic juice bottle and submerge it in the marsh, letting it fill with water. Slip a funnel into the mouth of the jar to aid in catching the gas. While still holding the jar underwater and keeping the funnel in place, turn the whole rig upside down and catch the gas bubbles that rise as you probe the mud with your shoe or a stick. Cap the jar tightly underwater, then bring it to the surface. Light a match and then carefully remove the lid and light the gas. It will burn blue for about thirty seconds. Do not use a larger container. A young boy I know once tried this experiment using a washtub and so created a small bomb.

FIGURE 6–26: A sample rig for catching marsh gas. (*Photo by Michael J. Caduto.*)

When bubbles break the surface of a marsh there may indeed be a creature lurking below, but it is probably just knocking loose a marsh gas pocket. Perhaps the mysteries of the murky marsh waters can best be remembered with this "Marsh Gas Ballad":

Who knows a feeling that can beat
the feel of mud beneath our feet?
It slips and slides between the toes
that walk among the cattail rows.
The bubbles that come floating up
live in the gushing, oozing glup.
And if they meet a lighted match
beware the blue, hot marsh-gas flash.
MICHAEL CADUTO

❧ADDITIONAL READING❧

Bent, Arthur Cleveland. *Life Histories of North American Birds.* Washington, D.C.: U.S. Government Printing Office. Multivolume collection, 1919-1958. Reprinted as a 26-volume set by Dover Publications, New York, 1961–1968.

Ernst, Carl H., and Roger W. Barbour. *Turtles of the United States.* Lexington, Ky.: University Press of Kentucky, 1972.

Errington, P. L., *Of Men and Marshes.* Ames, Iowa: The Iowa State University Press, 1969.

Godin, Alfred E., *Wild Mammals of New England.* Baltimore, Md.: The John Hopkins University Press, 1977.

Good, R.E., D.F. Whigham, and R.L. Simpson, eds., *Freshwater Wetlands: Ecological Processes and Management Potential.* New York: Academic Press, 1978.

Gore, A.J.P., *Mires: Swamp, Bog, Fen and Moor, Ecosystems of the World,* Vol. 4A, *General Studies.* New York: Elsevier Scientific Publishing Co., 1983.

Gore, A.J.P., *Mires: Swamp, Bog, Fen and Moor, Ecosystems of the World,* Vol. 4B, *Regional Studies.* New York: Elsevier Scientific Publishing Co., 1983.

Greeson, P.E., J.R. Clark, and J.E. Clark, *Wetland Functions and Values: The State of Our Understanding.* Minneapolis, Minn.: American Water Resources Association, 1979.

Horwitz, E. L., *Our Nation's Wetlands: An Interagency Task Force Report.* Washington, D.C.: Council on Environmental Quality, 1978.

Hotchkiss, Neil, *Common Marsh, Underwater and Floating-Leaved Plants of the United States and Canada.* New York: Dover, 1972.

Kusler, J.A., *Our National Wetland Heritage: A Protection Guidebook.* Washington, D.C.: Environmental Law Institute, 1983.

Larsen, J.A., *Ecology of Northern Lowland Bogs and Conifer Forests.* New York: Academic Press, 1982.

Magee, Dennis W., *Freshwater Wetlands: A Guide to Common Indicator Plants of the Northeast.* Amherst, Mass.: The University of Massachusetts Press, 1981.

Moore, P.D., and D.J. Bellamy, *Peatlands*. New York: Springer-Verlag, Inc., 1974.

Niering, William A., *The Life of the Marsh: The North American Wetlands*. New York: McGraw-Hill, 1966.

Shaw, S.P., and C.G. Fredine, *Wetlands of the United States*, Circ. 39.ʼ Washington, D.C.: United States Fish and Wildlife Service, 1956.

Thomas, Bill, *The Swamp*. New York: W.W. Norton and Co., Inc., 1976.

Ursin, Michael J., *Life in and Around Freshwater Wetlands*. New York: Thomas Y. Crowell Co., 1975.

﹡GENERAL ACKNOWLEDGMENT﹡

F. C. Golet was generous in providing major assistance in preparing this chapter. For general reference, I recommend the following:

F.C. Golet and J.S. Larson, *Classification of Freshwater Wetlands in the Glaciated Northeast* (Washington, D.C.: U.S. Fish and Wildlife Service Resource Publication 116, 1974).

This publication served as a general reference for many of the characteristics described for northeastern wetland types.

﹡NOTES﹡

1. W.E. Frayer, et al., *Status and Trends of Wetlands and Deepwater Habitats in the Coterminus United States, 1950's to 1970's* (Washington, D.C.: United States Fish and Wildlife Service, 1983). Available from: Department of Forest and Wood Sciences, Colorado State University, Fort Collins, CO.

2. L.M. Cowardin, et al., *Classification of Wetlands and Deepwater Habitats of the United States* (Washington, D.C.: U.S. Government Printing Office, 1979): 3.

3. R.G. Wetzel, *Limnology* (Philadelphia, PA: Saunders, 1975).

4. Cowardin et al., *Classification of Wetlands*.

5. P.V. Glob, *The Bog People* (London: Paladin, 1971) 142.

6. Ralph W. Tiner, Jr., *Wetlands of the United States: Current Status and Recent Trends* (Washington, D.C.: U.S. Government Printing Office, 1984).

7. Golet and Larson, *Classification of Freshwater Wetlands*.

8. C.H. Wharton, *Southern River Swamp: A Multiple Use Environment* (Atlanta: School of Business Administration, Georgia State University, 1970).

9. Tiner, *Wetlands of the United States*.

10. Ibid.

AFTERWORD

I invite you to travel, in your mind's eye, to a favorite retreat along the
edge of a pond, marsh, or river, and to reflect on these explorations
through the worlds of fresh water. During our journey we have heard the
plaintive call of the loon and witnessed the playful antics of otters. We have
seen the remarkable homes of determined and creative case-building caddisflies, symbolic of the natural order lying behind the seeming chaos that
meets our gaze across the face of a pond in summer, teeming with life. The
wonders of fresh water depend on the tiniest and least conspicuous plants and
animals for their survival—algae, the decomposers, and microscopic animals. All of this life rests in the hands of humankind.

Secrets of freshwater life reveal themselves best to an inquisitive and
discerning eye and a probing mind. A childlike sense of wonder will be
continually rewarded on its explorations. It is for this reason, perhaps, that a
pond is such a natural attraction to children of all ages. I would like to close
this journey with a childhood recollection and some reflections on water as
giver of life, as artist and sage.

Goose Pond lay in a shallow valley between two hills that were
rounded off ages ago by the glacier's scouring mass. It was a once-idyllic
setting of old pasture, mostly grown into young pole forest of oak, ash, and
hickory. Strangely, one thick stone wall had been laid, or so it seemed, right
through the pond, as if the farmer had used it as we did, for an extended
wharf from which to ambush frogs, turtles, and snakes. In truth, the valley
was once lined with drier turf.

One cold day during the in-between season of late winter and early
spring, I slipped away from the house, empty jar in hand, to that often-
frequented, ice-encrusted shore. This was a common visit to an old friend
whose many secrets I had mostly uncovered by now, or so I thought.

Perched on the half-submerged stone wall, I peered past the reflections on the water's surface and down into a sunny nook that was partly obscured by the shadow of a large granite rock. A tiny wisp of a creature darted from the murky shade and into the morning light. I moved quickly, determined not to let this fragile-looking pond creature become another evanescent mystery of the deep, often glimpsed yet never seen. In an instant the capture jar approached from one side, the lid from the other, and what once held dill pickles became the cage of my aquatic zoo.

"A shrimp, but not really a shrimp," was my first notion. Swimming upside down, it seemed to sink whenever it paused. Feathery feet beat in rhythmic waves, swishing its see-through body along. There was a tiny dark eye on each side of its head that seemed to look right through me. Could it see me? In trying to assess whether it had a state of mind, of consciousness, I drew the jar closer and stared intently at those eyes. I felt it then, while looking at that tiny life through the water that was stained lightly tea-brown from the fallen oak leaves. Something dissolved between us, a comfortable boundary I had always drawn and had assumed was not crossable. From the eyes of this tiny, shrimplike creature came an unnerving, questioning voice.

"Why have you caught me in this jar? I don't hurt you! What will you do with me? It's too hot in the sunlight, my eyes hurt. Who are you, anyhow? Where have you come from? You are strange to live out of water. My world is the pond. Please put me back."

But my curiosity had to be satisfied before I could release this tiny voice. For some inexplicable reason, I had to see it out of the water, in *my* element, before setting it free. I put a twig under the tiny shrimp and it rose from the pickle jar, draped like a wet rag. Deprived of the water's buoyancy, its body folded over the wood. A few quick seconds in the sunlight—at least

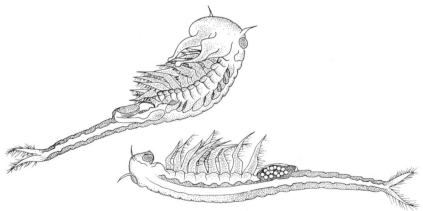

FIGURE A–1: Fairy shrimp.

FIGURE A–2: Ancient Carboniferous forest, home for some of the first land-dwelling animals.

they were quick to me—and one good look at the creature on my stick, and I laid its motionless body back into the jar. How cruel the acts of a curious mind can be: The fairy shrimp never made it home. Is a life too high a price to pay to satisfy a sense of wonder?

What is the nature of the bond between a young child, water, and a tiny freshwater crustacean? Here is shared a common spirit. It is an ancient lineage that harks back to the wet places, primitive and primeval, where sluggish beasts once wallowed in steamy sloughs amid giant ferns.

> *Dawn mist dew,*
> *stirring on sleepy pond.*
> *Breezes soft;*
> *lace-leaf wand*
> *conducting a toad chorus.*
>
> *Primeval pall*
> *Unearthly call*
>
> *Ancient voice from age of*
> *cave-bone tools.*
> *Call to daybreak*
> *life in murky pool.*

Over 300 million years ago, from the drying ooze along the edge of a sun-scorched pool, a fish gasped for air. Living a marginal existence in oxygen-poor pools of fresh water, it was the first of all air-breathing life to gulp air directly without the help of gills. In time—geologic time, that is—during a period that probably stretched longer than the time humans have since existed on earth, this half fish, half land dweller gradually became a creature that breathed with lungs.

Today there lives a paradox, belonging neither to water nor to land, that dwells in the interstices of the wet and dry. The mudskipper, *Periophthalmus*, draws air from water in its cheek pouches through a lunglike network of blood vessels. Living in the tidal flats of northern Australia, it climbs mangrove trees with its fins and catches insects. Or it skips along the mud flats on its pectoral fins, eating like a shorebird. If pursued, it can leap a foot high; it dodges predators by changing directions each time it touches down. This fish's eyes focus best in the air.

There are people who believe that an ancient teleost from the Paleozoic era, a *Crossopterygian*, is a common ancestor of all land life, including hominids. Whatever your convictions about the origin of human life, without water, life would not exist. It flows within and without all living things, impartial as to whether it moves along a riverbed or courses through the wings of a newly emerged butterfly, transforming them from a rumpled heap into gossamer tools of flight.

FIGURE A–3: Mudskipper, an acrobatic, paradoxical fish at home both on the land and in water.

FIGURE A–4: (*Photo by Michael J. Caduto.*)

FIGURE A–5: The view of Walden Pond from Thoreau's house site. (*Photo by Michael J. Caduto.*)

People ask what water is.
It is the origin of all things,
and the ancestral temple of all life.
Water produces the beautiful and the ugly,
the virtuous and wicked,
the foolish and the clever.

AUTHOR UNKNOWN,
5th century B.C., CHINA

We seek from water more than life itself. It is to the sea, or to ponds, lakes, and rivers that people often go to be inspired. There is a transcendent feeling in a moment spent listening to the breaking of waves along the shore, or watching the ever-changing ripples of a stream. Thoreau sensed this eternal current:

> While I bask in the sun on the shores of Walden Pond, by this heat and this rustle, I am absolved from all obligations to the past. The council of nations may reconsider their votes; the grating of a pebble annuls them.

Likewise Emerson:

> After raffling all day in Plutarch's morals, or shall I say angling there, for such fish as I might find, I sallied out this fine afternoon through the woods to Walden water. . . .
>
> As I sat on the bank of the Drop, or God's Pond, and saw the amplitude of the little water, what space, what verge, the little scudding fleets of ripples found to scatter and spread from side to side and take so much time to cross the pond, and saw how the water seemed made for the wind, and the wind for the water, dear playfellows for each other,—I said to my companion, "I declare this world is so beautiful that I can hardly believe it exists."

As varied and alive as a dancing candle's flame, the surface of water mesmerizes. It is a canvas that takes the image of the world above it. Ripples and ringlet waves animate images that may appear lifeless as stone.

reflected in a pool
sparkle of the morning dew
sunlight dancing

Yet this elixir of life is not merely a flowing canvas. Water is mentor, the consummate sage. In its soft, yielding example is strength that is seldom understood.

FIGURE A–6: (*Photo by Michael J. Caduto.*)

FIGURE A–7: (*Photo by Cecil B. Hoisington.*)

The highest good is like that of water.
The goodness of water is that it benefits the
ten thousand creatures, yet itself does not wrangle,
but is content with the places that all men disdain.
It is this that makes water so near to the Tao.

That the yielding conquers the resistant
and the soft conquers the hard
is a fact known by all men,
yet utilized by none.

LAO TZU, 5th century B.C.

FIGURE A–8: *(Photo by Michael J. Caduto.)*

Through the seasons, over the years, the movement of water is the ebbing and flowing of life itself.

windblown
racing, spinning, drifting
the leafboat passes by
a journey

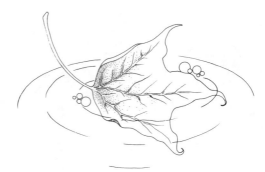

FIGURE A–9

Trickling or torrential, fluid or frozen, water is the element that reminds us of our place in the natural order. Life is inextricably linked to the seasonal interplay of water and sun, ice and north winds.

In early spring, before the last mound of snow has melted from the shade of a boulder, the underwater dance of the fairy shrimp begins. As they have done for millenia, the shrimp prepare fertile eggs for the later season when only mud will remain in their ephemeral ponds. Somewhere, a deft hand and inquisitive mind will capture one of these delicate creatures. The curious eyes of shrimp and human will meet and, perhaps, if both spirits are awake and listening, they will touch across an abyss of consciousness that has existed since the first glimmer of human thought.

FIGURE A–10: (*Photo by Cecil B. Hoisington.*)

APPENDIX A

Plant and Animal Adaptations for Aquatic Life

&GAS EXCHANGE&

Exchange of Gases Through External Tissues: Many amphibians, immature insects, and microscopic animals use this method. Submerged plants, too, can get carbon dioxide and oxygen directly from the water.

Gills: There are diverse forms of gills, many of which also serve to help the animal maintain a proper osmotic balance.

Physical gills consist of a bubble that is carried on the body, usually under the wings or on the abdomen. The bubble acts like an aqualung and supplies air to the submerged insect. Oxygen diffuses into the bubble as it is used up. Some insects have a thick *pile* of hairs under their abdomen which supports the air bubble, with as many hairs as two million per square millimeter. This is called *plastron* breathing. Beetles and true bugs commonly use physical gills.

Blood gills are an extension of the thin body cuticle through which gases diffuse. Mayflies often have elaborate blood gills on their abdomen.

Spiracular gills are canals found under a thin layer of skin into which oxygen diffuses. Black fly and crane fly larvae use this method.

Fish gills involve the flow of blood beneath a layer of skin only a few cells thick, into which air moves. Using a *counter current multiplier*, the blood flows toward the fresh water so it always contacts water containing higher levels of oxygen.

Hemoglobin: Hemoglobin, an iron compound, binds with oxygen and carries it throughout the body to individual cells. A midge larva, the blood worm, and many higher animals possess hemoglobin, which gives their blood a red color. Some others, like crayfish, use copper, resulting in blue blood.

Internal Lungs: These are found in many reptiles, amphibians and, of course, mammals, such as beavers and muskrats.

Habitat Preferences: Many species of stonefly and mayfly live only in fast-running, well-aerated water.

Air Spaces: Some plants store gases in their stems for nighttime use when photosynthesis is not possible. A mosquito larva, *Mansonia*, will pierce these tissues to obtain air.

Stomata: Gas exchange pores on the leaves of floating leaved aquatic plants are located on the top of leaves, as in water lilies, where the leaf is exposed directly to the air. Land plants have stomata on the undersides of their leaves where they are better protected in the terrestrial environment.

❧STAYING PUT IN THE CURRENT❧

Flattening to stay out of the current (mayflies, stoneflies).

Streamlining, such as in fish and of submerged leaves, to decrease resistance to the current (many pondweeds).

Small Size

Ballast is used by those caddisflies that create their homes of heavy sand grains.

Grasping Devices include suckers (leeches, snails), hooks, silk threads (black fly larvae), and jelly (snail egg masses).

Current Avoidance, such as hiding under rocks, moving into slower current, and burrowing into sediments.

❧MOVEMENT❧

Swimming: Fish swim with side-to-side motion; leeches undulate up and down as well.

Oar-Like Legs: found on many insects such as water boatmen and diving beetles. These "oars" are often formed by a broad surface area composed of stiff hairs.

Jet Propulsion: used by dragonfly nymphs as they shoot water from their anal pores.

Wriggling: mosquito and midge larvae.

Webbed Feet: beaver, muskrats, ducks, otters.

҈FINDING OPTIMAL TEMPERATURE҈

Endotherms, or warm-blooded animals, maintain a fairly constant body temperature independent of the environment.

Ectotherms, or cold-blooded animals, must rely mostly on the environment to provide body heat. They have evolved a number of ways of seeking to control their body temperature. Some are *stenothermic*, and have little tolerance for temperature changes. They usually live in environments with fairly constant temperatures such as cool, spring-fed streams. *Poikilothermic* animals have broad tolerance to temperatures. This is especially true for amphibians, frogs, salamanders, and newts, who will often sun themselves in order to increase their temperature and metabolic rate.

҈OSMOTIC BALANCE҈

Water tends to move through tissues from a solution where a dissolved substance is less concentrated to a solution where the substance is more concentrated. This process is called *osmosis*. Since the cellular fluids of freshwater animals are more concentrated than the water in their environment, they tend to take on too much water. This creates a need to constantly excrete excess water. Some mechanisms used to maintain a proper osmotic balance are:

Waterproof Body Surfaces, as found in many reptiles such as snakes and turtles, that allow the animal to control body fluids independent of their environment.

Water Excretory Organs: help to remove excess water. *Kidneys* are used by fish to prevent the dilution of body fluids, while retaining needed salts found in the water. *Contractile vacuoles*, small structures that fill with a substance and then empty into the surrounding environment, are used by some protozoans, such as *Paramecia* and *Amoeba*, to remove excess water. *Gills* often aid in osmotic regulation.

⛄RESPONSES TO CHANGES IN THE ENVIRONMENT⛄

Aquatic organisms, like their terrestrial relatives, must adapt to the extremes of winter cold and summer heat and drought.

Hibernation: The metabolic processes of hibernating animals decrease in response to shorter periods of daylight and colder temperatures. Virtually all amphibians hibernate under the mud or in nearby leaf litter, including frogs, toads, and salamanders, and most reptiles as well, such as snakes and turtles. Hibernation usually ends in the spring when days grow longer, temperatures rise, and water becomes more abundant.

Aestivation: Similar to hibernation, aestivation is a response to extreme heat and drought. Many amphibians aestivate during hot, dry summer periods.

Diurnation: This daily response to the stress of heat finds many animals least active from mid-afternoon until evening, seeking the cool, shaded areas of a pond or stream.

Reproductive Adaptation and Dispersal

Dormant Reproductive Stages: Some protozoans, crustaceans, and others form *resting eggs* or *cysts* that are resistant to drying and, in some cases, can hatch on being exposed to water after 20 years of dormancy. Bryozoans produce *statoblasts* that form an entire colony when spring arrives.

Asexual Reproduction among both plants and animals helps them to become quickly established in an environment. Root sprouts, called *rhizomes*, help buckbean to grow out over the water and extend the mat in a bog. Water willow has arching stems that serve a similar function. Water milfoil, bladderwort, and other plants form vegetative buds that can overwinter to produce new plants in the springtime. Many protozoans, like the *Amoeba*, can reproduce simply by dividing in half and forming new individuals.

Life Cycle Timing: Many species of animals emerge during the season and times of day when temperatures, food availability and current speed are optimal.

Dispersal: Frogs can hop to a new pond, and some water striders and other water bugs and beetles can fly to new areas. But many organisms have tiny structures—resting eggs, cysts, spores—that can be carried on the wind, washed downstream in debris, or transported in the digestive tracts of animals. Some wily critters, like leeches, actually travel attached to the feet of waterfowl or even on the back of a migrating turtle. Many microscopic invertebrates are so well dispersed that they are cosmopolitan, being distributed worldwide. Examples are some rotifers, sponges, protozoans, and crustaceans such as a waterflea, *Daphnia longispina.*

❧POLLINATION❧

Some aquatic plants are wind-pollinated as on land: grasses, sedges, rushes, and cattails. Others are pollinated by insects, such as water lilies and iris, while a few produce specialized pollen that is waterborne, like wild celery and hornwort.

❧ADAPTATIONS FOR EATING IN THE WATER❧

Many of the same feeding mechanisms are used by aquatic life as are used on land. One additional method is *filter feeding,* by which water is sieved for fine food particles. Freshwater mussels use this method in still waters, as do black fly larvae in streams.

❧ADAPTATIONS FOR LIVING IN FINE SEDIMENTS❧

Flattening helps to make many insects more streamlined while moving around in the muck.

Hairs can create a space between an insect and the surrounding sediments. These are common on stoneflies and mayflies.

Attaching to Larger Objects gives many small organisms a more stable environment.

Specialized Breathing Structures are often needed when sediments are in contact with an organism's body because there is danger of suffocation. Some insects have gills that are enclosed by other parts of their bodies. Dragonfly nymphs have internal, rectal gills. Still other insects have gills on their backs, where they are held above the sediments.

❧ADAPTATIONS IN PIGMENTATION❧

Animals that are exposed to high levels of light tend to be heavily pigmented. Pigments absorb light and protect underlying tissues from damage by the sun's harmful ultraviolet rays. Conversely, cave dwellers that live in total darkness are usually lacking in pigments, giving them an albino appearance.

Countershading, the lighter coloration on the bottom side of most animals, may help to offset the effects of shadows and cause the animal's outline to be less conspicuous against the colors of its environment. Also, when viewed from below, the animal will be less conspicuous against a light sky.

APPENDIX B

List of Common and Latin Names of Organisms Appearing in This Book

These organisms are arranged in general categories under each of the three Kingdoms of Monera, Plantae (Plants), and Animalia (Animals). They are listed alphabetically within each category, according to common names. The Latin names alone are given for those organisms that have no generally used common name. The major groupings are:

MONERA

Bacteria

PLANTS

Fungi
Algae
Dinoflagellates
Ferns
Mosses
Liverworts
Horsetails
Quillworts
Flowering Plants (Angiosperms) and Cone-bearing Plants (Conifers)
    Herbaceous Flowering Plants (plants lacking woody stems)
    Trees and Shrubs (the names of Conifers are followed by a "C")

ANIMALS

Invertebrates (animals without a backbone)
    Arthropods (insects, crustaceans, spiders, mites)
    Mollusks (snails, clams and mussels)
    Protozoans (single-celled animals)
    Sponges
    Hydras and Jellyfishes

Rotifers
Moss Animals (Bryozoans)
Worms
Water Bears (Tardigrades)
Vertebrates (animals with a backbone)
Reptiles and Amphibians
Fish
Birds
Mammals

MONERA

Bacteria
    Actinomycetes (phylum)
    Escherichia coli
    Leptothrix spp.
    Pseudomonas spp.
    Rhizobium spp.
    Sphaerotilus spp.

PLANTS

Fungi
    Hyphomycetes (phylum)
    Penicillium spp.
    Water Mold, *Saprolegnia ferax*

Algae
- Diatoms, Bacillariophyta (phylum)
    Asterionella spp.
    Cocconeis placentula
    Cyclotella stelligera
    Fragilaria spp.
    Melosira spp.
    Navicula radiosa
    Navicula spp.
    Nitzschia sigmoidea
    Tabellaria spp.

- Blue-green Algae, Cyanophyta (phylum) [often classified as Cyanobacteria (phylum)]
    Anabaena spp.
    Aphanizomenon spp.

238 / Appendix B

Microcystis spp.
Oscillatoria spp.
Rivularia spp.

- Golden Algae, Chrysophyta (phylum)
- Green Algae, Chlorophyta (phylum)
  Chlorella spp.
  Cladophora glomerata
  Euglena spp.
  Muskgrass, *Chara canescens*
  Muskgrasses, *Chara spp.*
  Nitella spp. (see Stoneworts)
  Oedogonium spp.
  Pediastrum spp.
  Phacus spp.
  Stoneworts, *Nitella spp.*
  Stonewort, *Nitella tenuisima*
  Trachelomonas spp.
  Ulothrix zonata
  Volvox spp.
  Zoochlorella (see Chlorella spp.)
- Green Algae, Gamophyta (phylum)
  Closterium spp.
  Micrasterias spp.
  Spirogyra spp.
- Red Algae, Rhodophyta (phylum)
  Hildenbrandia prototypus

Dinoflagellates, Dinoflagellata (phylum)
  Ceratium hirundinella

Ferns
  Fairy Moss (see Fern, Water)
  Ferns, Filicinophyta (phylum)
  Fern, Cinnamon, *Osmunda cinnamomea*
  Fern, Fragile, *Cystopteris fragilis*
  Fern, Ostrich, *Matteuccia struthiopteris*
  Fern, Royal, *Osmunda regalis*
  Fern, Sensitive, *Onoclea sensibilis*
  Fern, Water, *Azolla spp.*

Mosses
  Hypnum, Water, *Hygrohypnum ochraceum*
  Moss, Fountain, *Fontinalis spp.*
  Moss, Peat (see Moss, *Sphagnum*)

Moss, *Sphagnum, Sphagnum spp.*
Moss, Boat-leaved *Sphagnum, Sphagnum palustre*
Moss, Water, *Dichelyma capillaceum*
Mosses, Musci (class)
Water Hypnum (see Hypnum, Water)

Liverworts

Liverwort, *Jungermannia lanceolata*
Liverwort, *Ricciocarpus spp.*
Liverworts, Hepaticae (class)

Horsetails

Horsetail, *Equisetum spp.*
Scouring Rush (see Horsetail)

Quillworts

Quillwort, *Isoetes spp.*

Flowering Plants (Angiosperms) and Cone-bearing Plants (Conifers)

Herbaceous Flowering Plants (plants lacking woody stems)

American Lotus (see Lotus, American)
Arethusa, Arethusa bulbosa
Arrow Arum, *Peltandra virginica*
Arrowhead, *Sagittaria spp.*
Arrowhead, Broadleaf, *Sagittaria latifolia*
Bedstraw, Small, *Galium trifidum*
Bladderwort, Common, *Utricularia vulgaris*
Bladderwort, Horned, *Utricularia cornuta*
Buckbean, *Menyanthes trifoliata*
Burreed, *Sparganium spp.*
Calopogon, Calopogon pulchellus
Cattail, Broad-leaved, *Typha latifolia*
Cattail, Narrow-leaved, *Typha angustifolia*
Coltsfoot, *Tussilago farfara*
Coontail, *Ceratophyllum demersum*
Cowslip (see Marsh Marigold)
Dieffenbachia, Dieffenbachia spp.)
Duck Potato (see Arrowhead)
Duckweed, Great, *Spirodela spp.*
Duckweed, Small, *Lemna minor*
Elodea, Common, *Elodea canadensis*
Fanwort, *Cabomba caroliniana*
Grass Pink (see *Calopogon*)
Grasses, Poaceae (family), also called Gramineae (family)
Grass, Canary (see Grass, Reed Canary)

Grass, Cotton, *Eriophorum spp.*
Grass, Reed Canary, *Phalaris arundinacea*
Groundnut, *Apios americana*
Hornwort (see Coontail)
Iris, Iris spp.
Jewelweed, *Impatiens spp.*
Lady's Slipper, Yellow, *Cypripedium calceolus*
Lady's Tresses, *Spiranthes spp.*
Lotus, American, *Nelumbo lutea*
Marsh Marigold, *Caltha palustris*
Mermaid Weed, *Proserpinaca palustris*
Naiads, *Najas spp.*
Naiad, *Najas gracillima*
Orchis, White-fringed, *Habenaria blephariglottis*
Philodendron, Philodendron spp.
Pickerelweed, *Pontedaria cordata*
Pitcher Plant, *Sarracenia purpurea*
Pogonia, Rose, *Pogonia ophioglossoides*
Pondweed, Floating, *Potamogeton natans*
Pondweed, Sago, *Potamogeton pectinatus*
Rushes, Juncaceae (family)
Rush, Chairmaker's (see Threesquare)
Rush, Soft, *Juncus effusus*
Saint Johnswort, Marsh, *Hypericum virginicum*
Sedges, Cyperaceae (family)
Sedge, Three-way, *Dulichium arundinaceum*
Sedge, Tussock, *Carex stricta*
Skunk Cabbage, *Symplocarpus foetidus*
Smartweed, *Polygonum spp.*
Spanish Moss, *Tillandsia usneoides*
Spatterdock, *Nuphar luteum*
Spikerush, *Eleocharis spp.*
Sundew, *Drosera spp.*
Sundew, Thread-leaved, *Drosera filiformis*
Swamp Candles, *Lysimachia terrestris*
Tearthumb, *Polygonum sagittatum*
Threesquare, *Scirpus americanus*
Venus' Flytrap, *Dionaea muscipula*
Watercress, *Nasturtium officinale*
Water Hyacinth, *Eichornia crassipes*
Water Lily, White, *Nymphaea odorata*
Water Lily, Yellow (see Spatterdock)
Water Parsnip, *Sium suave*

Watermeal, *Wolffia spp.*
Watermilfoil, *Myriophyllum spp.*
Watermilfoil, Northern, *Myriophyllum exalbescens*
Watershield, *Brasenia schreberi*
Wild Celery, *Vallisneria americana*
Wild Rice, *Zizania aquatica*

Trees and Shrubs (the names of Conifers are followed by a "C")
Alder, *Alnus spp.*
Alder, Speckled, *Alnus rugosa*
Ash Black, *Fraxinus nigra*
Azalea, Swamp, *Rhododendron viscosum*
Bay, Loblolly, *Gordonia lasianthus*
Beech, American, *Fagus grandifolia*
Blueberry, *Vaccinium spp.*
Bog Rosemary (see Rosemary, Bog)
Buttonbush, *Cephalanthus occidentalis*
Cedar, Atlantic White, *Chamaecyparis thyoides* (C)
Cranberry, *Vaccinium spp.*
Cypress, Bald, *Taxodium distichum* (C)
Dogwood, Red Osier, *Cornus stolonifera*
Dogwood, Silky, *Cornus amomum*
Elderberry, *Sambucus spp.*
Fetterbush, *Lyonia lucida*
Gum, Black, *Nyssa sylvatica*
Gum, Tupelo, *Nyssa aquatica*
Labrador Tea, *Ledum groenlandicum*
Lambkill (see Laurel, Sheep)
Larch, *Larix laricina* (C)
Laurel, Bog, *Kalmia polifolia*
Laurel, Sheep, *Kalmia angustifolia*
Leatherleaf, *Chamaedaphne calyculata*
Mangrove, Black, *Avicennia germinans*
Mangrove, Red, *Rhizophora mangle*
Mangrove, White, *Laguncularia racemosa*
Maple, Red, *Acer rubrum*
Maple, Swamp (see Maple, Red)
Oak, *Quercus spp.*
Oak, Pin, *Quercus palustris*
Oak, Swamp White, *Quercus bicolor*
Oak, Water, *Quercus nigra*
Pine, Pond, *Pinus serotina* (C)
Pine, Swamp (see Pine, Pond) (C)

Rose, Swamp, *Rosa palustris*
Rosemary, Bog, *Andromeda glaucophylla*
Soapberry (see Sweet Pepperbush)
Spicebush, *Lindera benzoin*
Spruce, Black, *Picea mariana* (C)
Sweet Gale, *Myrica gale*
Sweet Pepperbush, *Clethra alnifolia*
Tamarack (see Larch) (C)
Titi, *Cyrilla racemiflora*
Wax Myrtle, *Myrica cerifera*
Willow, *Salix spp.*
Willow, Pussy, *Salix discolor*
Willow, Water, *Decodon verticillatus*
Winterberry, *Ilex verticillata*

ANIMALS

Invertebrates (animals without a backbone)
Arthropods (includes insects, crustaceans, spiders, mites)
Alderflies, Sialidae (family)
Alderfly, *Sialis spp.*
Backswimmers, Notonectidae (family)
Beetles, Coleoptera (order)
Beetles, Predaceous Diving, Dytiscidae (family)
Beetle, Predaceous Diving, *Dytiscus spp.*
Beetle, Riffle, *Psephenus herricki*
Beetle, Whirligig, *Dineutes spp.*
Bloodworm [see Midges (true)]
Bug, Giant Water, *Belostoma spp.*
Bug, Giant Water, *Lethocerus spp.*
Bug, Jesus (see Water Strider)
Bug, Ripple, *Rhagovelia spp.*
Bugs (true), Hemiptera (order)
Caddisflies, Trichoptera (order)
Caddisfly (free-swimming), *Rhyacophila spp.*
Caddisfly, Log Cabin, *Brachycentrus nigrosoma*
Caddisfly, *Neureclipsis spp.*
Caddisfly, Snail Case, *Helicopsyche spp.*
Copepods, Copepoda (subclass)
Crayfishes, Decapoda (order)
Crayfish, *Cambarus spp.*
Crayfish, *Orconectes spp.*
Crustaceans, Crustacea (class)

Cyclops, *Cyclops spp.*
Damselflies, Zygoptera (suborder)
Damselfly, Black-winged, *Calopteryx maculata*
Darning Needles (see Damselflies and Dragonflies)
Dobsonflies, Corydalidae (family)
Dobsonfly, *Corydalus spp.*
Dragonflies, Anisoptera (suborder)
Fireflies, Lampyridae (family)
Fishflies, Corydalidae (family)
Fishfly, Saw-horned, *Chauliodes serricornis*
Flies (true), Diptera (order)
Flies, Black, Simuliidae (family)
Flies, Crane, Tipulidae (family)
Flies, Soldier, Stratiomyidae (family)
Flies, Flower (see Rat-tailed Maggot)
Flies, Hover (see Rat-tailed Maggot)
Flies, Moth, Psychodidae (family)
Flies, Syrphid (see Rat-tailed Maggot)
Fly, Black, *Simulium venustum*
Fly, Crane, *Tipula spp.*
Fly, Soldier, *Odontomyia sineta*
Glowworm (see Fireflies)
Gnats, Buffalo (see Flies, Black)
Hellgrammite (see Dobsonflies)
Honeybee, *Apis mellifera*
Insects, Insecta (class)
Lightningbugs (see Fireflies)
Lily Leaf Caterpillar, *Nymphula spp.*
Lobster, *Homarus americanus*
Marsh Treaders (see Water Measurers)
Mayflies, Ephemeroptera (order)
Mayfly, *Baetis spp.*
Mayfly, *Ephemera varia*
Mayfly, *Ephemerella spp.*
Mayfly, *Hexagenia spp.*
Mayfly, *Stenonema spp.*
Meganeura spp. (dragonfly from the Carboniferous Period)
Midges, Net-winged, Blephariceridae (family)
Midge, Phantom, *Chaoborus spp.*
Midges, Phantom, Chaoborinae (subfamily)
Midges (true), Chironomidae (family)
Mites, Water, Arachnida (class)

Mosquito, *Culex spp.*
Mosquito, *Mansonia spp.*
Mosquito, *Wyeomyia smithii*
Mosquito Hawks (see Flies, Crane)
Mosquitos, Culicinae (subfamily)
Pillbugs (see Sowbugs, Aquatic)
Pond Skaters (see Water Striders)
Rat-Tailed Maggots, Syrphidae (family)
Scorpions, Arachnida (class)
Scuds, Amphipoda (order)
Shrimp, Fairy, Anostraca (order)
Shrimp, Seed, Ostracoda (subclass)
Sideswimmers (see Scuds)
Snowfleas, *Achorutes nivicolus*
Sowbugs, Aquatic, Isopoda (order)
Spiders, Arachnida (class)
Spider, Fish, *Dolomedes spp.*
Spider, Wolf, *Lycosa triton*
Spongillaflies, Sisyridae (family)
Springtails, Collembola (order)
Stoneflies, Plecoptera (order)
Stoneflies, Giant, Pteronarcidae (family)
Stoneflies, Winter, Taeniopterygidae (family)
Water Boatmen, Corixidae (family)
Water Fleas, Cladocera (order)
Water Flea, *Daphnia spp.*
Water Flea, *Daphnia longispina*
Water Measurers, Hydrometridae (family)
Water Penny (see Beetle, Riffle)
Water Scorpions, Nepidae (family)
Water Strider, *Gerris spp.*
Water Striders, Gerridae (family)
Water Strider, Broad-shouldered, *Rhagovelia spp.*
Water Striders, Broad-shouldered, Veliidae (family)
Water Treader, *Mesovelia spp.*
Water Treaders, Mesoveliidae (family)

Mollusks (snails, clams and mussels)
Clam, *Anodonta imbecillis*
Clams, Pelecypoda (class)
Clams, Fingernail, Sphaeridae (family)
Clams, Pill, Sphaeridae (family)
Clam, Razor, *Ensis directus*

Glochidium (see Mussels, Freshwater)

Limpet, *Lanx patelloides*

Mussels, Freshwater, Pelecypoda (class)

Slugs, Gastropoda (class)

Snails, Gastropoda (class)

Winkle, Gastropoda (class)

Protozoans (single-celled animals)

Amoeba spp.; in some classification systems as Rhizopoda (phylum)

Paramecium spp.; in some classification systems as Ciliophora (phylum)

Tokophrya spp.; in some classification systems as Ciliophora (phylum)

Vorticella spp.; in some classification systems as Ciliphora (phylum)

Sponges

Sponge, Freshwater, *Heteromeyenia tubisperma*

Sponge, Freshwater, *Spongilla lacustris*

Sponges, Porifera (phylum)

Hydras and Jellyfishes

Hydra, Green, *Chlorohydra viridissima*

Hydras, Coelenterata (phylum); in some classification systems as Cnidaria (phylum)
  Hydrozoa (class)

Jellyfish, Freshwater, *Craspedacusta sowerbyi*

Jellyfishes, Coelenterata (phylum); in some classification systems as Cnidaria (phylum)
  Scyphozoa (class)

Rotifers, Rotifera (phylum)

Bryozoans (Phylum), see Moss Animals

Moss Animals, Bryozoa (phylum); in some classification systems as Ectoprocta (phylum)

Moss Animal, *Pectinatella magnifica*

Worms

Bristleworm, *Nais spp.*

Earthworms, Aquatic, Annelida (phylum)
  Oligochaeta (class)

Flatworms, Platyhelminthes (phylum)

Leeches, Annelida (phylum)
  Hirudinea (class)

Nematodes, Nematoda (phylum)

Planaria, Planaria spp. (see Flatworms)

Roundworms (see Nematodes)

Worms, Gordian (see Worms, Horsehair)

Worms, Horsehair, Nematomorpha (phylum)

Worms, Proboscis, Nemertina (phylum)
Worms, Sludge (see Worms, *Tubifex*)
Worms, *Tubifex*, Annelida (phylum)
    Oligochaeta (class)

Tardigrades (see Water Bears)

Water Bears, Tardigrada (phylum)

Vertebrates (animals with a backbone)
Reptiles and Amphibians
Alligator, American, *Alligator mississippiensis*
Amphibians, Amphibia (class)
Bullfrog, *Rana catesbeiana*
Cottonmouth, *Agkistrodon piscivorous*
Crocodile, American, *Crocodylus acutus*
Frog, Green, *Rana clamitans*
Frog, Leopard, *Rana pipiens*
Frog, Meadow (see Frog, Leopard)
Frog, Mink, *Rana septentrionalis*
Frog, Pickerel, *Rana palustris*
Frog, Wood, *Rana sylvatica*
Mudpuppy, *Necturus maculosus*
Newt, Red-spotted, *Notopthalmus viridescens*
Red Eft (see Newt, Red-spotted)
Reptiles, Reptilia (class)
Salamanders, Urodela (order)
Salamander, Dusky, *Desmognathus fuscus*
Salamander, Spotted, *Ambystoma maculatum*
Salamander, Two-lined, *Eurycea bislineata*
Snake, Common Garter, *Thamnophis sirtalis*
Snake, Common Water, *Natrix sipedon*
Spring Peeper, *Hyla crucifer*
Toad, American, *Bufo americanus*
Toad, Fowler's, *Bufo woodhousei fowleri*
Turtle, Bog, *Clemmys muhlenbergii*
Turtle, Eastern Painted, *Chrysemys picta*
Turtle, Mud, *Kinosternon subrubrum*
Turtle, Stinkpot Musk, *Sternotherus odoratus*
Turtle, Red-eared, *Chrysemys scripta elegans*
Turtle, Snapping, *Chelydra serpentina*
Turtle, Spotted, *Clemmys guttata*
Fish
Alevin (see Salmon, Atlantic)
Alewife, *Alosa pseudoharengus*

Bass, Largemouth, *Micropterus salmoides*
Bluegill, *Lepomis macrochirus*
Bullhead, Brown, *Ictalurus nebulosus*
Carp, *Cyprinus carpio*
Catfish (see Bullhead, Brown)
Chub, Creek, *Semotilus atromaculatus*
Crappie, Black, *Pomoxis nigromaculatus*
Dace, Black-nosed, *Rhinicthys atratulus*
Eel, American, *Anguilla rostrata*
Flounder, Summer, *Paralichthys dentatus*
Flounder, Winter, *Pseudopleuronectes americanus*
Herring, Cisco (see Herring, Lake)
Herring, Lake, *Coregonus artedi*
Horned Pout (see Bullhead, Brown)
Killifish (see Mummichog)
Lamprey, Sea, *Petromyzon marinus*
Miller's Thumb (see Sculpin)
Minnows, Cyprinidae (family)
Minnow, Blunt-nosed, *Pimephales notatus*
Muddler (see Sculpin)
Mudskipper, *Periophthalmus koelreuteri*
Mummichog, *Fundulus heteroclitus*
Muskellunge, *Esox masquinongy*
Perch, Yellow, *Perca flavescens*
Pickerel, Chain, *Esox niger*
Pickerel, Redfin, *Esox americanus*
Pike, Northern, *Esox lucius*
Pike perch (see Walleye)
Pumpkinseed, *Lepomis gibbosus*
Salmon, Atlantic, *Salmo salar*
Salmon, Landlocked, *Salmo salar sebago*
Salmon, Sebago (see Salmon, Landlocked)
Sculpin, *Cottus cognatus*
Sculpins, Cottidae (family)
Shad, American, *Alosa sapidissima*
Shiner, Golden, *Notemigonus crysoleucas*
Smelt, *Osmerus mordax*
Stickleback, Brook, *Eucalia inconstans*
Stickleback, Three-spined, *Gasterosteus aculeatus*
Sturgeon, Lake, *Acipenser fulvescens*
Sucker, Common, *Catostomus commersoni*
Sunfishes, Centrarchidae (family); (see Bluegill; Crappie, Black; Pumpkinseed)

Trout, Brook, *Salvelinus fontinalis*
Trout, Brown, *Salmo trutta*
Trout, Lake, *Salvelinus namaycush*
Trout, Rainbow, *Salmo gairdneri*
Walleye, *Stizostedion vitreum*
Whitefish, Lake, *Coregonus clupeaformis*

Birds

Anhinga, *Anhinga anhinga*
Birds, Aves (class)
Blackbird, Red-winged, *Agelaius phoeniceus*
Bufflehead, *Bucephala albeola*
Coot, American, *Fulica americana*
Cormorants, Phalacrocoracidae (family)
Duck, American Black, *Anas rubripes*
Duck, Ring-necked, *Aythya collaris*
Duck, Ruddy, *Oxyura jamaicensis*
Duck, Wood, *Aix sponsa*
Eagle, Bald, *Haliaeetus leucocephalus*
Egret, Snowy, *Egretta thula*
Falcon, Peregrine, *Falco peregrinus*
Flycathers, Tyrannidae (family)
Gallinule, Purple, *Porphyrula martinica*
Goldeneye, Common, *Bucephala clangula*
Goose, Canada, *Branta canadensis*
Grebes, Podicipedidae (family)
Grebe, Eared, *Podiceps nigricollis*
Grouse, Ruffed, *Bonasa umbellus*
Gull, Herring, *Larus argentatus*
Gull, Ring-billed, *Larus delawarensis*
Harrier, Northern, *Circus cyaneus*
Hawk, Duck (see Falcon, Peregrine)
Hawk, Marsh (see Harrier, Northern)
Hawk, Red-shouldered, *Buteo lineatus*
Heron, Great Blue, *Ardea herodias*
Heron, Green-backed, *Butorides striatus*
Heron, Little Blue, *Egretta caerulea*
Killdeer, *Charadrius vociferus*
Kingbird, Eastern, *Tyrannus tyrannus*
Kingfisher, Belted, *Ceryle alcyon*
Kinglet, Golden-crowned, *Regulus satrapa*
Loon, Common, *Gavia immer*
Mallard, *Anas platyrhynchos*

Merganser, Common, *Mergus merganser*
Merganser, Hooded, *Lophodytes cucullatus*
Osprey, *Pandion haliaetus*
Owl, Barred, *Strix varia*
Owl, Great Horned, *Bubo virginianus*
Pheasant, Ring-necked, *Phasianus colchicus*
Pintail, Northern, *Anas acuta*
Rail, Virginia, *Rallus limicola*
Redhead, *Aythya americana*
Robin, American, *Turdus migratorius*
Sandpiper, Spotted, *Actitis macularia*
Scaup, Lesser, *Aythya affinis*
Shoveler, Northern, *Anas clypeata*
Snipe, Common, *Gallinago gallinago*
Sparrow, Song, *Melospiza melodia*
Sparrow, Swamp, *Melospiza georgiana*
Sparrow, White-throated, *Zonotrichia albicollis*
Swallow, Tree, *Tachycineta bicolor*
Swallows, Hirundinidae (family)
Swans, Cygninae (subfamily)
Teal, Blue-winged, *Anas discors*
Thrushes, Turdidae (family)
Turkey, Wild, *Meleagris gallopavo*
Vireos, Vireonidae (family)
Warbler, Magnolia, *Dendroica magnolia*
Warbler, Myrtle (see Warbler, Yellow-rumped)
Warbler, Palm, *Dendroica palmarum*
Warbler, Northern Parula, *Parula americana*
Warbler, Yellow-rumped, *Dendroica coronata*
Warblers, Wood, Parulidae (family)
Waterthrush, Louisiana, *Seiurus motacilla*
Waterthrush, Northern, *Seiurus noveboracensis*
Waxwing, Cedar, *Bombycilla cedrorum*
Woodcock, American, *Scolopax minor*
Woodpecker, Pileated, *Dryocopus pileatus*
Wren, Long-billed Marsh (see Wren, Marsh)
Wren, Marsh, *Cistothorus palustris*
Yellowlegs, Lesser, *Tringa flavipes*
Yellowthroat, Common, *Geothlypis trichas*

Mammals
Bat, Little Brown, *Myotis lucifugus*
Bear, Black, *Ursus americanus*

Beaver, *Castor canadensis*
Deer, White-tailed, *Odocoileus virginianus*
Fox, Red, *Vulpes fulva*
Lemming, Southern Bog, *Synaptomys cooperi*
Mammals, Mammalia (class)
Manatee, Florida, *Trichechus manatus*
Mink, *Mustela vison*
Mole, Star-nosed, *Condylura cristata*
Moose, *Alces alces*
Mouse, Meadow Jumping, *Zapus hudsonius*
Mouse, White-footed, *Peromyscus leucopus*
Muskrat, *Ondatra zibethicus*
Myotis, Little Brown (see Bat, Little Brown)
Nutria, *Myocastor coypus*
Opossum, *Didelphis marsupialis*
Otter, River, *Lutra canadensis*
Rabbit, Cottontail, *Sylvilagus spp.*
Rabbit, Marsh, *Sylvilagus palustris*
Raccoon, *Procyon lotor*
Shrew, Masked, *Sorex cinereus*
Shrew, Water, *Sorex palustris*
Vole, Red-backed, *Clethrionomys gapperi*

GLOSSARY

Acid deposition. *See* Acid precipitation.

Acid precipitation. A form of air pollution consisting of all forms of precipitation, including wet and dry particles, that have an acidity lower than normal rainfall (pH 5.6).

Acid rain. *See* Acid precipitation.

Adaptation. A genetic or behavioral trait or pattern which enhances a plant's or animal's ability to survive and reproduce in its environment.

Adsorption. The adhesion of a very thin layer of molecules to the surfaces of solids or liquids with which they come into contact.

Adult. A mature organism (usually a plant or animal) that is fully developed and able to reproduce.

Aeration, zone of. A place within the soil and around the bedrock where the pore space is not saturated with water.

Aerobic. (1) Occurring, acting, or living only in the presence of oxygen; (2) pertaining to organisms needing oxygen to live.

Aestivation. A period of dormancy similar to hibernation, during which an animal retreats from the heat of the summer and experiences a lower body temperature and reduction of growth and metabolic processes.

Alkalinity. The total of all compounds contained in water that have the ability to combine with or neutralize acids, causing a shift in pH toward the alkaline side of neutrality. A measure of buffering capacity.

Anadromous. Fish that live in salt water and return to fresh water to spawn.

Anaerobic. (1) In the absence of oxygen; (2) pertaining to organisms capable of respiration in the absence of oxygen.

Angiosperms. "Vessels for seeds." Flowering plants. The class Angiospermae, within the division Tracheophyta.

Animalia. The animal kingdom, including animals with and without a backbone. Under the five-kingdom classification system, the kingdom Animalia is defined as including all organisms that develop from a blastula (a type of embryo).

Antagonism. When two or more actions or agents work against each other, decreasing the effectiveness of all.

Aphotic. Lacking light. *See* Tropholytic.

Aquifer. A water-bearing layer of permeable rock, sand, or gravel.

Aufwuchs. *See* Periphyton.

Autotrophs. Organisms that can create organic nutrients from inorganic substances by using the sun's energy via photosynthesis or through the oxidation of inorganic compounds. *See* also Producers, primary.

Battery. Large floating mats of peat and vegetation, buoyed by gases formed from decomposition of peat during low-water periods. Gases cause the organic materials to become buoyant and float free when the wet season comes.

Bedrock. Solid rock that underlies the soil and fragmented rock.

Benthic. Located on the bottom of a body of water or in the bottom sediments or pertaining to bottom-dwelling organisms.

Benthos. Organisms that live on the bottom of aquatic environments, such as lakes and ponds.

Biological half-life. The time it takes for a body to rid itself of one half its load of a substance.

Biological oxygen demand (BOD). A measure of the amount of oxygen-demanding decomposition and respiration required to fully consume the organic matter contained in a given sample or body of water.

Biomass. The total weight or volume of all living things found in a defined location.

Bog. A wetland formed where surface drainage is congested. Low oxygen levels and soil temperatures cause incomplete decomposition, resulting in the buildup of fibrous peat. Only specialized plants can grow in these extreme conditions. Mosses (especially *Sphagnum spp.*), sedges, and lichens can tolerate bog conditions; one or more of these groups form the dominant plant community in a bog. The water and soil of *Sphagnum* bogs are typically highly acidic, which further inhibits decomposition.

Boundary layer. The 1–3-millimeter (.04–.1 inch) space between flowing water and the substrate, in which friction between the water and bottom causes a decrease of current velocity to zero where the water meets the bottom. The boundary layer is an important dwelling place for benthic animals, plants and microbes because it provides relative safety from being washed downstream.

Brackish. Water of mild salinity, usually where fresh water and salt water meet as a river empties into the ocean.

Bryophytes. One of two major groups in the plant kingdom, bryophytes are distinguised from the tracheophytes by the lack of fluid-conducting tissues of xylem and phloem, and the production of motile sperm that must swim through water to fertilize eggs. The conspicuous generation is the gametophyte. The three main classes are the liverworts (Hepaticae), mosses (Musci), and hornworts (Anthocerotae).

Budding. Reproduction occurring asexually through a small outgrowth or bud that pinches off from the parent organism to form a new individual.

Buffering capacity. The ability of a solution to resist changes in pH when acidic or basic substances are added.

Capillary zone. *See* Aeration, zone of.

Carboniferous. Geologic time period that stretched from 345 million years ago to 280 million years ago.

Carnivore. One that eats animals.

Catadromous. Fish that live in fresh water and return to the ocean to spawn.

Cephalothorax. The forward part of the body on some arthropods, consisting of a fused head and thorax.

Cerci. Tail appendages found on some insects, usually long and thin or clasperlike. These possibly serve a role in reproduction.

Chemosynthetic. *See* Chemotrophic.

Chemotrophic. Pertaining to bacteria that can synthesize organic compounds and derive energy from inorganic elements through chemical reactions, without the aid of sunlight.

Cilium. (1) A short, hairlike, usually numerous structure of microscopic size used by cells to provide locomotion or to move fluid and small particles across the cell or into the mouth area. Cilia are sometimes called *undulipodia*. (2) Sometimes used loosely as a synonym for flagellum. *See* Flagellum.

Climax community. A stage during ecological succession that is relatively stable and during which plants and animals are capable of continued reproduction in the environment they have helped to create. Climax communities can be altered by disruptions such as storms, climatic changes, disease, fire, or human activity.

Commensalism. A symbiotic interaction between two organisms by which one benefits and the other is unaffected.

Community. All the living things that dwell interdependently in a particular place and share the available energy and resources.

Compensation depth. The vertical zone of a lake where photosynthesis balances respiration. The boundary between the upper trophogenic (photosynthetic) zone and the lower tropholytic zone, where respiration and decomposition predominate.

Competition. Rivalry for the same limited resource(s) by two or more individuals or groups of individuals.

Condensation. The changing of a substance from a gas to a liquid, usually as a result of cooling.

Conifers. Cone-bearing plants. The class Coniferae within the division Tracheophyta.

Consumers. Organisms that cannot make their own food and must obtain energy by eating other living things.

Consumers, microbial. Mites, nematodes, springtails, and other organisms that consume microbial algae, bacteria and fungi. These form an essential link in the process of decomposition.

Countercurrent heat conservation. A circulatory system in which cold venal blood coming from the extremities is in close contact with warm arterial blood coming from the heart. This warms the venal blood and helps to maintain a constant body temperature.

Covalent bond. Chemical bonds formed by the sharing of pairs of electrons between the bonded atoms.

Crepuscular. Active at dusk or dawn.

Cycles, biogeochemical. The full circulation of biologically important nutrients or elements between the living and nonliving parts of the environment.

Cycle, gas. The cyclical exchange of an element between (1) the stores of that element found primarily in the atmosphere and (2) the living plants, animals, and microbes for which that element fulfills a nutrient or other requirement.

Cycle, sedimentary. The cyclical exchange of an element between (1) the stores of that element found primarily in rock, soil, and in solution in aquatic systems (e.g., phosphorus), and (2) the living plants, animals, and microbes for which that element fulfills a mineral or nutrient requirement.

Cycle, water. The continuous global circulation of water via evaporation from the earth's surface by the sun's heat; the rising, cooling, and precipitation of water back to earth; and the eventual evaporation of that water once again.

Cyst. A reproductive structure (usually a dormant stage) which is encapsulated and is resistant to drought, heat, and cold. Cysts form during times of environmental stress.

Decomposers. Organisms (such as bacteria and fungi) that break down plant and animal remains into forms once again usable by producers.

Deep-water habitat. Aquatic habitats, such as lakes, rivers, and oceans, where the water is deeper than 2 meters (6.6 feet), which is the deepest water in which emergent plants can grow. Water, not air, is the primary habitat in the deep waters.

Delta. A deposit of sand, silt, and clay where swift waters enter a slower body of water and drop their sediment load. Deltas are usually triangular in shape, with the triangle pointing upstream, toward the source of swifter water.

Denitrification. The changing of biologically usable nitrogen compounds (for example, nitrates [NO_3^-]) into atmospheric nitrogen (N_2) by fungi and bacteria.

Detritus. Dead plant, animal, and other organic matter.

Dimictic. Pertaining to temperate lakes, and those of colder, high elevations in subtropical regions, that experience a thermal mixing or overturn twice each year, usually in the fall and spring. *See* Overturn, spring and fall.

Discharge of streams and rivers. The volume of water passing a certain point along a stream or river in a given period of time.

Discharge of ground water. Pertaining to ground water that emerges at the surface, leaving the ground-water stores.

Dispersal. The movement of plants and animals into new habitats and locations. Dispersal can occur via the travels of adults, juveniles, young, or via the transport (by wind, water, animals, or other means) of specialized structures such as seeds, spores, eggs, and cysts.

Dissolved oxygen (DO). Oxygen dissolved in water.

Dissolved substances. Substances that have become chemically combined with a liquid such that the two can no longer be distinguished. Dissolved substances will not settle out of solution.

Diurnal. Of the daytime or occurring during the daylight hours.

Diurnation. A daily period of partial dormancy when an animal experiences a lowered body temperature and a reduction of growth and metabolic processes. Typical of bats during the daytime in temperate regions.

Diversity. The variety, number and distribution of species within a community.

Dorsal. Located on or near the back.

Drainage divide. A ridge of high land separating the areas that drain into different river systems or watersheds.

Drift. Algae, bacteria, detritus, and invertebrates that are carried downstream by the current.

Drift, stratified. *See* Stratified drift.

Dun. *See* Subimago.

Dune swale. A hollow or depression that has been formed among sand dunes by the sculpting action of wind.

Ecology. The study of the interactions between living things and their environment.

Ecosystem. Broadly, all the living and nonliving things found in a given area, such as plants, animals, sunlight, water, and soil.

Ecotone. The juncture of two or more different kinds of ecosystems. A place that possesses the qualities of several types of ecosystems.

Ectotherm. An animal that depends on the environment as its source of heat and which cannot regulate its body temperature independently of the environment. A cold-blooded animal.

Eddies. A kind of contrary turbulence that creates circular upstream currents behind rocks and other obstructions and along the edges of a stream or river channel. There is also a vertical movement of water in eddies, which mixes oxygen from above into the deeper layers, and provides thermal mixing.

Egg. Female reproductive cell whose nucleus contains one set of chromosomes. In reptiles, birds, and some other groups, eggs are structures in which embryonic development occurs, and usually consist of an external shell, membranes, and an egg cell surrounded by nourishment (yolk).

Egg, resting. Eggs produced by fertilization among small and microscopic animals such as water fleas and rotifers. These eggs have a thick shell and are resistant to environmental extremes of drought, cold, and heat. They are capable of overwintering and are often produced during the fall and times of environmental stress.

Emergents. Rooted plants that can tolerate flooded soil but not extended periods of being completely submerged.

Endangered species. Species whose surviving numbers have dropped to such extremely low levels that they are in immediate danger of extinction. Human destruction of habitat and outright killing of plants and animals has frequently caused the numbers of surviving individuals within a species to decline to the endangered status.

Endotherm. An animal that produces its own body heat internally and is capable of regulating its body temperature independently of the environment. A warm-blooded animal.

Energy flow. The passing of energy along a food chain between living things in an ecosystem. Some energy is lost, due to growth and maintenance, as the energy passes between trophic levels.

Epilimnion. The upper thermal layer that forms in lakes during the summer when warmer, less dense water is buoyed by the colder water below. *See* Metalimnion and Hypolimnion.

Erosion. The removal or wearing away of soil or rock by water, wind, or other agents.

Estuarine. Being of, like, or from estuaries.

Estuary. Place where fresh water and salt water meet at the mouth of a river or where coastal freshwater springs are upwelling under a tidal zone.

Eutrophic. Referring to a productive body of water, high in organic matter and mineral nutrients (e.g., phosphate and nitrate) and often exhibiting seasonal oxygen deficiency.

Eutrophication. The overfertilization of aquatic ecosystems, resulting in high levels of production and decomposition. Eutrophication can hasten the aging process of a pond or lake due to the rapid buildup of organic remains.

Evaporation. The changing of a substance from a liquid to a gas by exposure to the air and/or heat.

Exoskeleton. An external skeleton, as in insects.

Facultative. The condition of being optional, not obligate. Capable of living with or without something, such as oxygen.

Filter feeder. Organism that feeds by sieving fine food particles (plants, animals, or detritus) from the water.

Flagellum. (1) A whiplike structure on the surface of a cell that is used for locomotion or to move fluids and small particles across the cell or into the mouth area. Flagella are composed of *flagellin,* a protein. They are usually long and few in number. (2) Also used loosely as a synonym for cilium. *See* Cilium.

Floodplain. Broad, flat lands along a river or stream that normally become inundated during floods, resulting in the deposition of sediments.

Food chain. A way of showing how nutrients and energy pass from producers through the various trophic levels in an ecosystem, such as from producers to herbivores, carnivores, and finally decomposers.

Food pyramid. A graphic representation showing all the energy and biomass contained in each trophic level of an ecosystem at any given time, moving from producers up the food web to top-level consumers.

Food web. An integration of the many food chains existing in an ecosystem, showing the complex, interwoven pathways of energy flow between the organisms living in that environment.

Food web, dynamic. A food web that shows the quantity and types of food used by each organism represented in the food web.

Frond. The entire leafy part of a fern stalk, from root to tip.

Fry. Young fish.

Fungi. A division of non-photosynthetic, heterotrophic plants (e.g., mushrooms and molds) under the three-kingdom classification system. Fungi are a separate kingdom under the five-kingdom classification system.

Frustule. A diatom shell, high in silica.

Gametophyte. The gamete stage of a plant that has only a single set of each type of chromosome (haploid) and which is capable of producing gametes (eggs and sperm).

Gill, physical. An air bubble used by many insects, especially bugs and beetles, as a lunglike structure for breathing underwater. Oxygen diffuses into the air bubble as it is used up by the insect.

Gill. An out-folding of an animal's body wall that functions in gas exchange with the surrounding environment, usually water.

Glacial deposits. Rocks, gravel, sand, and other material carried by glaciers, or their meltwaters, and left behind when the glaciers melted. *See* Chapter 1 for specific types of deposits.

Glochidium. The larva of a freshwater mussel which parasitizes the gills and scales of a fish during approximately the larva's first thirty days of life.

Habitat. The kind of environment in which a certain organism normally lives.

Herbivore. A plant eater.

Heterotrophs. Organisms (consumers and decomposers) that need to ingest organic food created by autotrophs in order to survive.

Hibernation. A period of dormancy in animals during which there is a marked drop in the rate of metabolism and growth, including heartbeat, respiratory rate, circulation, and body temperature.

Homeothermy. *See* Endotherm.

Hydrogen bond. A weak chemical bond formed between one atom of hydrogen and two other highly electronegative atoms (e.g. oxygen, nitrogen, and fluorine). This bond is important in determining the physical and chemical properties of water.

Hydrology. The branch of science that studies the distribution of the earth's waters, above and below ground, including the water cycle.

Hypha. The threadlike, branching part of a fungus.

Hypolimnion. The lower, colder thermal layer that forms in lakes during the summer when the warmer, less dense water rises toward the surface, leaving the colder, denser water below. *See* Epilimnion and Metalimnion.

Ice. The solid, crystalline form of water that exists below the freezing point at sea level of 32°F (0°C).

Impermeable. Pertaining to a substance that does not allow fluids to penetrate.

Interglacial. A period of relative warmth worldwide when continental glaciers melt back to extreme northern latitudes.

Invertebrates. A popular term designating an animal without a backbone.

Isostatic rebound. The rise in elevation or rebounding of the earth's crustal plates. This followed the melting back of the glacier's massive volume and weight, which had depressed the earth's surface to lower elevations.

Kettle hole. A depression that is often steep-sided and bowl-shaped, that formed when an ice block (left buried in front of the melting glacier) melted and left a hole.

Kidney. An excretory organ, found among higher vertebrates, that separates water and waste products from the blood, producing urine, which is then excreted.

Lacustrine. Something that is of, like, or originating in lakes.

Lake, delta. A lake that forms at a river delta where sediment builds up, forming dams and levees that isolate a body of water. *See* Delta.

Lake, plunge pool. A lake that forms in the hole scoured at the base of a large waterfall.

Larva. Immature stage of some animals' development that usually undergoes major changes before changing into an adult form.

Lateral line. A long, thin sensory organ found on fish that runs on each side of the body from the head to the tail region. It senses currents and pressure changes.

Lentic. Of still waters.

Limiting element. An element essential for life that is in short supply. By adding quantities of a limiting element to an ecosystem, the plants and/or animals present will demonstrate a major, positive growth response.

Limnetic. Pertaining to the open-water or pelagic zone of a lake which lies beyond the littoral zone, and which extends down to the depth at which light levels are at least 1 percent of the available sunlight at the surface. *See* Littoral.

Limnology. The branch of aquatic ecology that studies freshwater environments and the life within.

Littoral. The shoreline zone of a lake where sunlight penetrates to the bottom and is sufficient to support rooted plant growth. *See* Limnetic.

Lotic. Of flowing water.

Macronutrients. Nutrients needed in relatively large amounts, such as phosphorus, calcium, magnesium, potassium, sulfur, chlorine, and sodium.

Macrophyte. A large plant, as opposed to small and microscopic plants such as algae.

Marine. Being of, like, or from the ocean.

Marl. *See* Travertine.

Marsh, freshwater. A wetland where standing water exists year-round, except in the shallower areas during late summer or unusually dry years. The deep-water limit occurs where the water exceeds 2 meters (6.6 feet) in depth. (*See* Deep-water habitat.) Marshes may support the growth of emergent plants such as cattails, bulrushes, reeds, sedges, pickerelweed, arrowhead, or arrow arum; floating-leaved plants like water lilies and pondweeds; floating plants such as duckweeds; and submergents. Soil may be sand, silt, or soft, black muck.

Meadow, wet. Seasonally flooded wetland where standing water is usually present during the late fall, winter, and early spring, and where the water table often drops to below the surface during the summer and early fall. Some meadows, such as those along a river floodplain, may only be flooded intermittently. Grasses, rushes, and sedges are dominant plants; soils range from black and mucky to being fairly well decomposed.

Mesotrophic. Of moderate fertility and productivity. Between oligotrophic and eutrophic.

Metalimnion. A thermal layer that forms in lakes during the warm season, intermediate between the upper warm layer (epilimnion) and the colder layer beneath (hypolimnion).

Metamorphosis, complete. Insects that undergo the four distinct developmental stages of egg, larva, pupa, and adult.

Metamorphosis, gradual. *See* Metamorphosis, incomplete.

Metamorphosis, incomplete. The mode of development of some insects during which the egg hatches into a nymph or naiad. The nymph looks like a small adult and it grows gradually larger, molting its exoskeleton, until it is reproductively mature.

Microhabitat. Habitats or homes on a very small scale, such as the underside of a leaf.

Micronutrients. Nutrients needed in relatively small or trace amounts, such as zinc, manganese, magnesium, silica, iron, and iodine.

Migration. The periodic movement of animals from one area to another and back again.

Mineralization. The process of becoming mineral, specifically the release of nutrients (minerals) contained in microbes and detritus, through leaching from detritus, excretion, and upon death of decomposers.

Monera. The kingdom that includes bacteria.

Monomictic. Lakes in warm climates that experience a constant thermal mixing only once each year, during the coldest season.

Moraine, end. Hills of unsorted glacial debris (till) that accumulated at the leading edge of the glacier when the rate of melting equaled forward movement.

Moraine, ground. Unsorted glacial debris (till) that originated from within and beneath the glacial ice. Ground moraine results in a landscape of gently rolling sags and swells.

Moraine, terminal. *See* Moraine, end.

Mother of pearl. A pearly, hard, protective coating that lines the shells of some mollusks.

Mutualism. A symbiotic interaction between two organisms from which both benefit.

Naiad. *See* Nymph.

Neuston. The community of plants and animals that live on top of, or suspended from, the surface film.

Niche. The ecological role or position that a living thing or group of living things occupies in an ecosystem. For instance, a fungus occupies the niche of decomposer when it breaks down a dead plant or animal.

Nitrogen fixation. The changing of atmospheric nitrogen (N_2) into nitrogen compounds (for example, nitrates [NO_3^-]) usable by living things. Fixation occurs biologically via bacteria and blue-green algae, and abiotically by lightning fixation and other means.

Nocturnal. Of the nighttime or active at night.

Nutrient spiral. The cycle of production and decomposition in a stream. Nutrients are produced primarily in runs and ripple areas, and decomposed primarily in pools and still backwaters, creating a spirallike flow of nutrients moving downstream.

Nymph. The interim stage of development, between egg and adult, among insects that undergo incomplete metamorphosis. Sometimes called a naiad.

Obligate. The condition of needing something to survive—for example, oxygen.

Oligotrophic. Low in nutrients and productivity.

Omnivore. An organism that eats several kinds of food, possibly including plants, animals, and detritus.

Operculum. Hard plate that closes the opening of a snail's shell to provide protection and prevent dessication. Made of fingernaillike material.

Organic matter. Elements or material containing carbon, a basic component of all living things.

Osmosis. The movement of a liquid through a semipermeable membrane such as living tissue. During osmosis, a liquid moves from a solution in which a dissolved substance is less concentrated, into one in which the substance is more concentrated.

Overturn, spring and fall. A period when thermal currents and wind stir up a pond or lake, mixing the water thoroughly from top to bottom. Important nutrient supplies are churned into the water from the bottom sediments, and oxygen is mixed in from above and from zones where green plants create oxygen through photosynthesis.

Oxbow pond or lake. A crescent-shaped body of water formed as a meandering river gradually cuts through the outer bank at the beginning of a looping turn and joins directly with the channel downstream from the turn. The water no longer flows through the oxbow, and a pond or lake is formed, depending on the size and depth of the abandoned river channel.

Oxygen sag. The drop in the oxygen content of water to anaerobic or nearly anaerobic levels. Associated with the high levels of aerobic respiration and oxygen consuming decomposition accompanying organic pollution.

Paleolimnology. The science that studies the remains of plants and animals in such habitats as lakes, bogs, and marshes. Diatom frustules, pollen grains, insects, and other animal remains are preserved in these areas of poor decomposition. These remains are used by paleolimnologists to reconstruct a picture of past plant and animal life and environmental conditions.

Palustrine. Something that is of, like, or originating in wetlands such as marshes, meadows, swamps, or bogs.

Parasitism. An interaction between two organisms from which one benefits and the other is harmed. Parasitism is usually to the detriment, but not death, of one organism.

Parthenogenesis. Reproduction without fertilization of the egg.

Parts per million (ppm). A unit of relative volume used to measure very small quantities. One part per million is equivalent to one liter placed in a lake with the surface area of a football field and a depth of 8.8 inches (22.4 centimeters).

Parts per million (ppm). A unit of relative volume used to measure very small quantities. One part per million is equivalent to one liter placed in a lake with the surface area of a football field and a depth of 8.8 inches (22.4 centimeters).

Peat. Soil composed of fibrous, spongy, partially decomposed organic matter. Peat forms under conditions where decomposition is incomplete, such as in wetlands. Peat is the typical soil of bogs.

Pelagic. *See* Limnetic.

Periphyton. Algae that are attached to substrates or to living things.

Permafrost. Permanently frozen soil.

pH. A measure of hydrogen ion concentration or acidity. The pH scale ranges from 0 (most acid) to 14 (most basic), with a pH of 7 being neutral. This scale is logarithmic, so a drop or rise of one whole number indicates a tenfold increase or decrease of acidity. A change of two whole numbers indicates a hundredfold change.

Photic zone. *See* Trophogenic.

Photosynthesis. The process during which green plants use chlorophyll, sunlight, water, and carbon dioxide to create water, oxygen, and carbohydrates such as starches, sugars, and waxes. Photosynthesis is the primary source of energy in the global ecosystem.

Phytoplankton. Plankton that is composed of tiny plants and plant matter which consists largely of algae. These plants are major sources of production in aquatic systems. *See* Zooplankton.

Plankton. Small plants, animals, or microbes that are free-swimming, floating or drifting in the open water, as opposed to being attached to a substrate. *Planktonic*.

Plant zonation. The zones of plants having different growth forms, which can be observed as a gradation moving from shallow to deep water (or vice versa): emergent zone, floating-leaved zone, submergent zone, and open water zone.

Plantae. The plant kingdom, including ferns, mosses, cone-bearing plants, flowering plants, algae, and fungi under the three-kingdom classification system (which is used in this book). Under the five-kingdom classification system, algae are members of the protoctista and fungi are a separate kingdom.

Plastron. The ventral or lower shell of a turtle.

Pleistocene. Geologic time epoch that stretched from roughly 2 million years ago to 10,000 years ago.

Pocosin. The name describing upland evergreen shrub swamps and forests in the coastal southeastern United States. Bay and pond pine are important trees, with leatherleaf, wax myrtle, fetterbush, and titi growing in the understory. *See* Wetland, scrub-shrub.

Pollination. The transfer of pollen from the male anther of a flower to the female stigma.

Pollution, chemical. The introduction of toxic substances into an ecosystem.

Pollution, ecological. (1) Adding a substance that is not naturally occurring into an ecosystem, or (2) increasing the amount or intensity of a naturally occurring substance, or (3) altering the level or concentration of a biological or physical component of an ecosystem. Pollution occurs when one of these three parameters stresses the organisms in an ecosystem and requires responses beyond their normal range of resiliency, causing sickness or death.

Pollution, organic. Oversupplying an aquatic ecosystem with nutrients and organic materials. This results in high rates of production and decay, with associated ecological problems, such as oxygen depletion.

Pollution, thermal. Substances introduced into an aquatic ecosystem at temperatures above or below the normal temperatures found there.

Porosity. The amount of space or pores within soil or rock, relative to the total volume of that substance.

Prairie potholes. Inland marshes of south-central Canada and north-central United States. These marshes have formed in millions of shallow potholelike depressions that were formed by the glaciers. Prairie potholes are the most valuable inland wetlands for waterfowl production in North America.

Precipitation. (1) Minute particles that fall to the earth, such as water (rain, snow, hail, sleet) and dust. (2) The separation of a substance from solution or suspension by chemical or physical change.

Predation. An interaction during which one organism (predator) kills and consumes another (prey). Predation is usually used to describe an interaction between two animals, but herbivores can also be considered predators upon plants, and some specialized plants are predaceous on animals.

Producers. *See* Producers, primary.

Producers, primary. Green plants and other autotrophs that are capable of changing inorganic elements into organic tissues (food energy).

Production, secondary. The production of living tissue (biomass) by the consumption of autotrophs.

Productivity. A measure of the ability of an ecosystem to sustain life, including such factors as fertility, climatic conditions, and the available sunlight and water.

Profundal. *See* Tropholytic.

Protoctista. The kingdom that includes protists, algae, slime molds, and other life forms (under the five-kingdom classification system).

Protozoa. Traditionally, under the three-kingdom classification system, a subkingdom of the Animalia, which has designated single-celled organisms such as *Amoeba, Paramecium,* some flagellates, and others as animals. These organisms are classified, under the five-kingdom system, as protoctists in the kingdom Protoctista.

Pseudopodia. "False feet." The leading extensions of a cell membrane and cytoplasm, among single-celled organisms, that are used for feeding, locomotion, and other functions. *See* Streaming.

Pupa. A usually inactive stage in an insect's development, during which it is usually enclosed in a protective case while it undergoes major anatomical changes before emerging as an adult.

Radula. A toothed, rasplike plate used by snails to scrape food into smaller particles.

Raft. A large assemblage of water birds on a body of water.

Recharge. When water enters the ground-water stores.

Reflection. The return of light, heat, or sound waves from the surface that they have struck.

Refraction. The bending of light, heat, or sound waves as they pass at an angle from one substance to another of a different density, through which they travel at a different speed.

Respiration. The process during which organisms metabolize food molecules to get energy for growth and maintenance, consuming oxygen and giving off carbon dioxide.

Riparian. Located or living along the edge of a stream or river.

Riverine. Something that is of, like, or originating in rivers.

Runoff. *See* Surface runoff.

Salt wedge. A wedge-shaped mass of tidal salt water that intrudes the mouth and lower course of a river. The denser salt water underlies the lighter, fresh water of the river.

Saprophyte. Plants that obtain their energy and nutrients from dead organic matter.

Saturation, zone of. Place lying below the water table where the pore space is saturated with water under pressure greater than that of the atmosphere. Ground water.

Scavenger. Animal that eats dead organic matter.

Scroll pond or lake. An elongated body of water formed on the inside of the bend of a river. The channel gradually erodes the outer bank and deposits sediment on the inside of the bend. This results in a slow migration of the channel and the abandonment and isolation of the old channel, which becomes a scroll pond or lake, depending on its size and depth.

Seed. Reproductive structure of a plant, comprised of an embryo and its food surrounded by a protective coat.

Seral stage. One of the transitional communities that becomes established during the process of ecological succession. *See* Succession, ecological.

Sere. In ecological succession: The whole sequence of ecological communities that succeed each other in the development from the early, "pioneer" stages of succession to the climax community.

Sessile. (1) Attached directly at the base, not free to move about (animal), or (2) lacking a stalk; directly attached to the main stem (plant).

Shredder. Animals that eat plant remains and break them down into smaller pieces.

Sinkhole. Depression that has formed where water dissolved a void in the bedrock and caused the overlying soil and rock to sink.

Spate. A sudden flood, sometimes called a flash flood.

Species. A population of organisms that is capable of reproducing among their own kind, but not with members of other, genetically dissimilar or geographically isolated groups. The concept of species is not strictly applicable to bacteria since widely different groups may trade genes.

Sperm. Male reproductive cell whose nucleus contains one set of each type of chromosome. Must fuse with an egg to form a new individual with two sets of chromosomes.

Spiracle. On insects, the opening of a trachea (breathing tube) through which gas exchange occurs with the environment.

Spore. Dust-sized or microscopic reproductive structure.

Sporophyte. A plant or life cycle stage of a plant that contains two sets of each type of chromosome and which is capable of producing spores.

Spring. Place where ground water flows to the surface.

Stigma. (1) A spot found on the wings of some insects, located near the tip along the leading edge. Formed by a thickening of the wing membrane. (2) The red eyespot of some algae. (3) A flower part found among angiosperms: The part of the female pistil that is receptive to the pollen (sperm).

Stomata. Small or microscopic openings found on the surface layer (epidermis) of cells of plant leaves or stems, through which gas exchange occurs with the surrounding environment. *Guard cells* surround each stomata and control its size, and thus the rate of gas exchange. This term is also used to refer to the entire structure, consisting of the guard cells plus their pores.

Stratified drift. Layers of glacial sediment deposited by meltwaters, in which the particles of each layer are of a fairly uniform size. Swifter waters deposit larger particles (sand, gravel, and rocks); slower waters deposit finer particles (silt and clay).

Stream, ephemeral. A stream that flows over a highly porous substrate and feeds or recharges the ground water. Ephemeral streams are short-lived, existing only after heavy rainfalls and drying up in between.

Stream, intermittent. Streams that feed or recharge the ground water. They flow only during the wet seasons of spring and early summer (when the water table is high) and after heavy rains during the rest of the year.

Stream, perennial. A stream that normally flows year-round because it lies at or below the ground-water table, which constantly replenishes it.

Streaming. A form of movement among single-celled organisms such as *Amoebas*. Through fluid motion within the cell membrane, extensions of the cell push out in one direction (*see* Pseudopodia), and the trailing edge of the cell is retracted to cause a change in location.

Subimago. A subadult stage among mayflies. The nymph emerges from the water and hatches into the subimago, which is dull-looking and sparsely covered with downlike hairs, and then molts once again into the adult. Also known as a dun.

Submergent. Plant that grows and reproduces while completely submerged.

Succession, ecological. The progressive change in structure and species composition of a community, caused by the gradual alteration of the environment by plants and animals that live there.

Surface film. The interface between water and the atmosphere above it, which, because of the force of the surface tension, forms a surface capable of supporting the weight of small and microscopic organisms. *See* Surface tension.

Surface runoff. Water flowing over the surface of the land.

Surface tension. The force at the water's surface that resists being penetrated by small and/or microscopic particles or living things. Surface tension is caused by the powerful attraction between water molecules, which is stronger at the surface than the attraction between water molecules and the air above them. *See* Surface film.

Suspended particles. Particles that are buoyed by the movement of water but will settle out when that movement slows or stops.

Swamp. Wetland where the soil is saturated and often inundated and trees (woody plants 20 feet [6.1 meters]) tall or taller form the dominant cover. Soils are rich in organic matter. Shrubs typically form a second layer beneath the forest canopy, with a layer of herbaceous plants growing beneath the shrubs.

Symbiotic. The relationship of two or more organisms of different species or type living together and interacting.

Synergism. A combination of effects of some actions or agents that are greater than the sum of the individual effects.

Tectonic. Having to do with forces and changes in the earth's crust.

Thermal stratification. The formation of vertical temperature zones in a lake or pond.

Thermocline. *See* Metalimnion.

Threatened species. Species whose surviving numbers are so low that they would become threatened with extinction if they declined further. Human destruction of habitat and outright killing of plants and animals have frequently caused species to become threatened.

Till. Unsorted heaps of rock, sand, and gravel deposited by glacial ice.

Tolerance levels. The concentrations of toxic elements which, if exceeded, will cause stress, lowered resistance to disease, and other environmental hazards, and will eventually result in the death of the affected organism(s).

Toxic element. An element that damages the ability of living things to carry out essential life functions, and disrupts vital ecological processes, such as nutrient cycling and energy flow.

Tracheophytes. One of two major groups in the plant kingdom, distinguished from bryophytes by the presence of fluid-conducting tissues xylem and phloem. The sporophyte forms the conspicuous generation. Tracheophytes include ferns, horsetails, clubmosses, cone-bearing plants, and flowering plants.

Travertine. Deposits of calcium carbonate found on the plants and bottom substrates of hard waters, particularly those rich in limestone.

Trophic level. The feeding level that an organism occupies in the food chain. For instance, carnivores can be first order (eating herbivores), second order (eating other carnivores that eat herbivores), or third order (eating carnivores that eat other carnivores).

Trophogenic. "Nutrient producing." Pertaining to the upper zone of a lake in which light levels are at least 1 percent of the available sunlight at the surface. The light supports the photosynthetic production of organic matter of this zone. Also called the photic zone.

Tropholytic. "Nutrient decomposition." Pertaining to the depths of a lake where light levels are less than 1 percent of the available sunlight at the surface. Photosynthesis cannot be supported here; decomposition is predominant. Also called the aphotic or profundal zone.

Turbidity. A measure of the ability of light to penetrate water, indicating the amount of living and nonliving suspended solids and dissolved substances.

Vacuole. A vessel or space in a cell that is surrounded by a membrane and usually filled with fluid.

Vacuoles, contractile. Specialized vacuoles that excrete water and other wastes from the cells of single-celled and simple multicellular organisms. The vacuoles expand as the wastes accumulate, then they contract and expel the wastes from the cell. *See* Vacuole.

Vacuole, food. A vacuole that digests food. *See* Vacuole.

Valve (molluskan). One of the halves or parts that make up the whole shell of a mollusk.

Velocity of flowing water. The distance traveled by water over a given period of time.

Velocity, critical. The velocity of water above which a plant or animal, on the stream or river bottom, will be washed downstream.

Ventral. Located on or near the belly or front.

Vernal pond. A temperate pond that exists only during the wet season of spring and dries up later, in the growing season.

Vertebrates. Animals with a backbone.

Viscosity. The friction within a fluid, and between that fluid and other substances, which causes the liquid to resist flowing and creates friction and drag when something tries to move through the liquid. Water's viscosity is caused by its tendency to form hydrogen bonds with neighboring molecules.

Water. The molecule composed of one atom of oxygen bonded to two atoms of hydrogen.

Water, fresh. Water that is not salt water. Clean, unpolluted water.

Water, ground. Water found at and beneath the water table, in the zones of soil and bedrock, which are saturated.

Water, hard. Water with a high content of dissolved mineral salts. Hard water is usually associated with soft, easily eroded rocks such as limestone.

Water, soft. Water that is low in dissolved mineral salts. Soft water is associated with hard, erosion-resistant rocks.

Water table. The uppermost level of the ground water, separating the zone of aeration (above) from the zone of saturation or ground water (below).

Water vapor. The gaseous state of water.

Watershed. The area drained by a river system.

Weathering. The breaking down of rock, soil, and other materials by exposure to sun, wind, rain, and other conditions, both natural and of human origin, such as acid rain.

Well, artesian. A well that taps ground water under sufficient pressure to create a natural flow to the surface.

Wetland. Transitional zones between dry land and deep water where "saturation with water is the dominant factor determining the nature of soil development and the types of plant and animal communities living in the soil and on its surface." (Definition from Cowardin, M., et al., *Classification of Wetlands and Deepwater Habitats of the United States* [Washington, D.C.: United States Government Printing Office, 1979] p. 2.)

Wetland, forested. *See* Swamp.

Wetland, moss-lichen. A wetland where mosses or lichens form the dominant plant cover in an area where the soil is composed of saturated peat. *See* Bog.

Wetland, scrub-shrub. Wetlands dominated by shrubs, woody plants that are less than 20 feet (6.1 meters) tall. Also called shrub swamps and pocosins. Water levels in shrub swamps can range from permanently flooded to being flooded intermittently.

Zooplankton. Plankton that is composed of tiny animals and animal matter (for example, eggs, larvae, and other immature stages of insects and crustaceans). *See* Phytoplankton.

INDEX

Big burreed, 73
Biogeochemical cycles, 29
Biological half-life of toxic elements, 39–40
Biological oxygen demand (BOD), 39
Biomagnification, 39–40
Biotic plants, 23
Bird nests, searching for, 95
Birds:
  bogs, 204
  changes in lake depth and, 22
  Latin and common names of, 248–49
  of marshes, 191
  of lakeshore, pond, and marsh, 94
  of open water of lakes, 131–34
  of ponds, 93–95
  seasons and, 107
  of stream environments, 170
  of swamps, 194–97
Bitterns, 93
Black ash, 197
Black bears, 170
Black crappie, 89
Black ducks, 204, 93
Black fly, 155, 161
Black fly larvae, 161, 170, 234
Black gum, 197
Black ice, 108
Black spruce, 203
Black-nosed dace, 169
Black-winged damselfly, 164
Bladderwort, 79, 201
Blood gills, 230
Bloodsucking leech, 100
Bloodworm, 99, 100, 127, 148, 172
Blue-green algae, 63, 64, 97
  blooms of, 125
  of hot springs, 156
  of lakes, 123, 124, 125
  of rivers, 171
  of springs, 156
Blue-winged teal, 94
Blueberry, 197, 210
Blunt-nosed minnow, 169
Boat-leaved sphagnum, 68
Bog animals, 203–205
Bog laurel, 203
Bog plants, 200–203
Bog rosemary, 203
Bog turtle, 204, 206
Bogs, 199–205
  decomposition in, 211
  ecological succession in, 202
  formation of, 199, 201
Bottom-dwelling pond animals, 97–105
Bottom environments:
  invertebrates as indicators of pollution of, 37
  of muddy river, 172–173
  of stream or river, 150, 151
Boundary layer, 153
Brachycentrus, 166
Brachycentrus nigrosoma, 166
Brasenia schreberi, 75
Breathing structures, adaptation to living in fine
  sediments, and, 234–35
Bristleworm, 160
Broad-leaved cattail, 72, 73
Broad-shouldered water strider, 159
Broadleaf arrowhead, 74
Brook stickleback, 169, 170
Brown bullhead, 90
Bryophytes, 65
Bryozoans, 81, 98–99, 125, 233
  Latin and common names of, 245
Buckbean, 201
Budding, 76
Buffalo gnat. See Black fly
Buffering capacity, 34
Bufflehead, 132

Bullfrog, 82, 90, 91, 193
Bullhead, 107
Burreed, 71–72
Buttonbush, 71, 72

C

Caddisfly, 81, 154, 164–66
Caddisfly larvae, 103, 148, 155, 166, 170
Calanoids, 85
Calcium, living conditions in ponds and, 59
Calcium carbonate deposits, in flowing waters, 147
Calopogon, 201
Cambarus, 167
Canada goose, 94
Canals, ocean fish in lakes and, 130
Canoeing:
  on lakes, 114–15
  on wetlands, 183–85
Capillary zone, 17
Carbon-14, radiocarbon dating and, 211
Carbon dioxide:
  algae and, 63
  ecosystem and, 28–31
  in ponds, 54
  respiration and, 25
Carboniferous forest, ancient, 222
Carnivores, 26
Carp, 89–90, 105, 173
Case-builders, 166
Catadromous fish, 131
Catfish, 105, 173
Cattails, 71–72, 172, 191, 192, 209
Cedar waxwing, 94, 204
Cellulose, decomposition by fungi, 61
Cephalanthus occidentalis, 72
Cephalothorax, 160
Ceratium hirundinella, 125
Cerci, 162–63
Chain pickerel, 90, 190
Chairmaker's rush. See Threesquare
Chamaedaphne calyculata, 71
Channelization, ecological damage and, 43
Chaoborus spp., 82
Chara, 65
Charles River Watershed, Mass., 213–14
Chauliodes serricornis, 165
Chemical testing kits, 111
Chemosynthetic bacteria, 123
Chemotrophic bacteria, 31
Chironomid midge larvae, 148
Chlorella spp., 97, 123, 124
Chlorohydra viridissima, 127
Chordata, 82
Cilia, 86
Cinnamon fern, 68, 69, 197
Cladocera, 85
Cladophora glomerata, 157
Clam, 81, 173, 175
Classification of living things, 61–62
Climate:
  adaptations to changes in, 233–34
  nutrient levels and, 35
  wetlands and, 209
Climax community, 79
Closterium spp., 63, 65
Clouds, water cycle and, 13
Cocconeis placentula, 157
Coleoptera, 81, 84
Coliform bacteria, 61
Collecting. See Sampling and collecting
Collecting pans, 111
Colonies of algae, 62
Coltsfoot, 157
Commensalism, 46
Common elodea, 66
Common garter snake, 92

Common goldeneye, 132
Common loon, 132
Common mergansers, 93
Common sucker, 173
Common water snake, 92
Common yellowthroat, 204
Community, 23
  diversity of, 24
  survey in wetland, 215–17
Compensation depth, 120
Competition, freshwater life and, 47
Complete metamorphosis, 81
Cone-bearing plants, Latin and
  common names of, 239
Conifers. See Cone-bearing plants
Consumers, 26
Contractile vacuoles, 232–33
Convergent evolution, 158
Coontail, 79
Cooperation, freshwater life and, 47
Coots, 93
Copepod, 85, 126–27, 128
Coriolis force, 118
Corydalus spp., 165
Cotton grass, 201
Cottonmouth, 92
Cottontail rabbit, 193
Countercurrent heat conservation, 131
Countercurrent multiplier, 230
Countershading, 235
Covalent bonds, 9
Cowslips, 194
Cranberry, 201, 210
Crane fly, 161, 162
Crane fly larvae, 154
Craspedacusta sowerbyi, 86
Crayfish, 101, 102, 167
Creek chub, 169
Critical velocity of water, 153
Crossopterygian, 223
Crustaceans:
  of hot springs, 156
  of pond bottoms, 101–102
  of river channel, 172
  of streams, 167
  of zooplankton, 85
Culex spp., 82, 83
Current, 143
  adaptations for staying put in, 231
  of flowing waters, 149–50
  of lakes, 118
Cyanobacteria, 64
Cyclopoids, 85
Cyclops, 85
Cyclotella stelligera, 171
Cypress "knees," 196, 197
Cyprinidae, 170
Cysts, of protozoans, 97–98

D

Dabbling ducks, 193–94
Dams:
  beaver, 95
  bottom-draining, 144
  ecological pollution and, 42, 44–45
  fish and, 44–45
  formation of lakes and, 117–18
  river temperature and, 145
  top-draining, 145
Damselfly, 164
Damselfly nymph, 103, 104, 141–43
Daphnia longispina, 234
Daphnia spp., 85, 128
DDT, peregrine falcons and, 198
Decapods, 101
Decomposers, Decomposition, 31